the LOST *species*

the

LOST

species

*Great Expeditions in the Collections
of Natural History Museums*

CHRISTOPHER KEMP

THE UNIVERSITY OF CHICAGO PRESS
CHICAGO AND LONDON

The University of Chicago Press, Chicago 60637
The University of Chicago Press, Ltd., London
© 2017 by Christopher Kemp
Published 2017
Paperback edition 2020
Printed in the United States of America

29 28 27 26 25 24 23 22 21 20   1 2 3 4 5

ISBN-13: 978-0-226-38621-8 (cloth)
ISBN-13: 978-0-226-51370-6 (paper)
ISBN-13: 978-0-226-38635-5 (e-book)
DOI: https://doi.org/10.7208/chicago/9780226386355.001.0001

Library of Congress Cataloging-in-Publication Data
Names: Kemp, Christopher, author.
Title: The lost species : great expeditions in the collections
of natural history museums / Christopher Kemp.
Description: Chicago ; London : The University of Chicago Press, 2017. |
Includes bibliographical references and index.
Identifiers: LCCN 2017016333 | ISBN 9780226386218 (cloth : alk. paper) |
ISBN 9780226386355 (e-book)
Subjects: LCSH: Natural history—Popular works. | Type specimens (Natural
history)—Popular works. | Biological specimens—Identification—Popular
works. | Natural history—Research—Popular works. | Natural history—
Catalogs and collections. | Natural history museums.
Classification: LCC QH45.5 .K46 2018 | DDC 508.075—dc23
LC record available at https://lccn.loc.gov

For my family

Tax·on·o·my

    takˈsänəmē/

    noun Biology

    the branch of science concerned with classification, especially
    of organisms; systematics

    1. the classification of something, especially organisms: "the
    taxonomy of these fossils"

    2. a scheme of classification

plural noun: taxonomies

"a taxonomy of smells"

    ORIGIN: early nineteenth century: coined in French from the
    Greek *taxis*, meaning "arrangement" and *nomia*, meaning
    "distribution"

The beginning of wisdom is to call things by their proper names.
Confucius

Collections are large-scale research facilities. The problem is, they
don't look like it. If you go to the Large Hadron Collider or if you
go and look at some other particle accelerator or a radio telescope
or something, you go in and there's lights and there's switches and
there's things that make noises. It's all gleaming and clean, and
there are people running around doing stuff on computers. It looks
very high-tech. If you into the basement of a museum and you see
a bunch of jars on shelves, you don't necessarily say, "That's in the
same category of things as a particle accelerator."

    But it is.
Christopher Norris, Yale Peabody Museum of Natural History

Nomina sunt consequentia rerum.
(Names are the consequences of things.)
Dante

CONTENTS

INTRODUCTION   xiii

The Vertebrates

1. Pushed up a Mountain and into the Clouds:
   The Olinguito (*Bassaricyon neblina*)   3

2. Beneath a Color 83 Sky: The Ucucha Mouse
   (*Thomasomys ucucha*)   13

3. Going on a Tapir Hunt: The Little Black
   Tapir (*Tapirus kabomani*)   19

4. A Taxonomic Confusion: The Saki
   Monkeys (*Pithecia* genus)   31

5. Scattered to the Corners of the World: The Arfak
   Pygmy Bandicoot (*Microperoryctes aplini*)   41

6. The One That Got Away for 160 Years: Wallace's
   Pike Cichlid (*Crenicichla monicae*)   51

7. Here Be Dragons: The Ruby Seadragon
   (*Phyllopteryx dewysea*)   59

8. A Century in a Jar: The *Thorius* Salamanders   67

9. From a Green Bowl: The Overlooked Squeaker
Frog (*Arthroleptis kutogundua*)   75

10. A Body and a Disembodied Tail: Smith's
Hidden Gecko (*Cyrtodactylus celatus*)   85

The Invertebrates

11. Treasure in the By-Catch: The Gall
Wasps (*Cynipoidea* species)   95

12. The Biomimic: The Lightning Cockroach
(*Lucihormetica luckae*)   103

13. Sunk beneath the Surface in a Sea of Beetles:
Darwin's Rove Beetle (*Darwinilus sedarisi*)   111

14. The Spoils of a Distant War: The Congo Duskhawker
Dragonfly (*Gynacantha congolica*)   119

15. A Specimen in Two Halves: Muir's Wedge-Shaped
Beetle (*Rhipidocyrtus muiri*)   127

16. Mary Kingsley's Longhorn Beetle
(*Pseudictator kingsleyae*)   135

17. The Giant Flies (*Gauromydas papavero* and
Gauromydas mateus)   143

18. It Came from Area 51: The Atomic Tarantula
Spider (*Aphonopelma atomicum*)   151

19. The Host with the Most: The Nematode
Worm (*Ohbayashinema aspeira*)   161

20. From a Time Machine on Cromwell Road:
Ablett's Land Snail (*Pseudopomatias abletti*)   169

21. In Sight of Land: Payden's Isopod
(*Exosphaeroma paydenae*)   177

22. A Ball of Spines: Makarov's King Crab
(*Paralomis makarovi*)   185

Botanical

23. In an Ikea Bag: The Custard Apple Family
(*Monanthotaxis* Genus)   193

The Others

24. Waiting with Their Jackets On:
The Fossils (Paleontology Specimens Collected
by Elmer Riggs)   201

25. The First Art: The Earliest Hominin
Engraving (a 500,000-Year-Old Shell)   211

Epilogue   217

ILLUSTRATION CAPTIONS AND CREDITS   223
NOTES   227   INDEX   237

INTRODUCTION

A couple of years ago, University of Tennessee entomologist Stylia-nos Chatzimanolis borrowed a box of unclassified beetles from the Natural History Museum in London. Taxonomists do this all the time; they borrow material from other natural history collections the way someone might borrow a book from a library. The beetles Chatzimanolis borrowed were unknown—unsorted. They'd been collected in the past by field biologists but never formally described.

When the box arrived in Chattanooga, Tennessee, Chatzimano-lis found twenty or so beetle specimens pinned inside it. But one of the beetles was not like the others. First, it looked much older. Long-bodied, with a segmented, sinuous abdomen, it was a rove beetle—but an unusually large specimen with a wide iridescent green head.

The beetle had been collected in Argentina in 1832. As Chat-zimanolis began to look more closely at the specimen—and at its yellowed handwritten label—he realized it had been collected by a young Charles Darwin during the voyage of the *Beagle*. Somehow it had never been described. It was stored away unnamed, then it had disappeared without a trace into the vast beetle collection in London.

Finally, after its 180 years in storage, Chatzimanolis gave the beetle a name. It is so unlike any known species of rove beetle that he had to erect an entirely new genus to contain it. He named it *Darwinilus sedarisi.*[1]

The beetle is named now, but the central question remains: Does it matter? Is there any need to name it at all? The specimen had waited for 180 years. Why name it now? In fact, it matters a great deal.

A single species is the irreducible component of all the biodiversity on Earth. For centuries scientists have been trying to describe, classify, and order the natural world. When a new species is named, an array of other work suddenly becomes possible. By studying a newly named species alongside its closest relatives, biologists gain a deeper understanding of the evolutionary processes that shaped it. Ecologists are given a window into the workings of infinitely complex ecosystems. Conservationists gain insights into how to manage environments to maintain its population numbers. But for all this to be possible we first need to know it exists.

We can think of Earth's biodiversity as a symphony, with each species represented by a single musical note. By itself, a single note means nothing—it is stripped of meaning. But over time notes accumulate and begin to intermingle. Notes become motifs, which become themes. Themes repeat and slowly build to become movements—rich musical passages that have become more than the sum of their parts. But what if the sheet music is incomplete? Currently we have named an estimated one-fifth of all species on Earth. Imagine what Beethoven's Ninth Symphony might sound like if an orchestra played only every fifth note and left silent gaps between them. What happens to the themes and the slowly gathering refrains? How can we begin to understand Earth's biodiversity in all its richness without knowing even half of its components?

Museums and biorepositories worldwide are filled with unknown species. By some estimates, 75 percent of newly described mammal species are already part of a natural history collection somewhere in the world. It's true for all other orders too: for parasitic worms and for frogs, fishes, corals, and flies; for crabs, moths, and lichens and bryophytes. In 2012 researchers discovered an unknown species of egg-eating sea snake at the Natural History Museum in Copenhagen.[2] Collected in the late 1800s, the snake was subsequently named *Aipysurus mosaicus*, for the mosaic-like pattern made by its brown and cream-colored scales. A long time ago it had been misidentified as a closely related species, labeled as such, then stored in a jar for more than a century. But it was different, and markedly so—different enough that a sharp-eyed herpetologist immediately noticed.

The authors of a 2010 *Proceedings of the National Academy of Science* study estimate that of the seventy thousand or so species of flowering plants waiting to be described, about half have already been collected.[3] They are stored without names in herbaria. It's not difficult to imagine how this might happen. The encyclopedic Department of Entomology at the National Museum of Natural History in Washington, DC, alone contains about thirty million insect specimens. There is far too much material for taxonomists and curators to assess, identify, and name.

Even when a specimen is examined, it's often misidentified and wrongly named. In a November 2015 article in *Current Biology*, researchers assessed the accuracy of collections in herbaria in twenty-one countries.[4] Essentially, more than half of all the specimens they examined had been given the wrong name: collections are filled with errors. This is a huge problem. How useful is a collection if the labels on the specimens are wrong? What if half the specimens aren't what we think they are? Between 1970 and 2000 the number of tropical plant specimens in herbaria doubled, but most of the recently collected material was mislabeled. And in plenty of cases specimens are not identified at all. A faded handwritten label on the specimen might bear the tantalizing annotation "novel species?" Or simply "nov. sp.?"

This is all made possible by how little we know about the natural world that thrums and vibrates all around us. We have barely scratched its surface. By most estimates, there are about ten million species on the planet, but we have named fewer than two million of them. The rest of Earth's biodiversity remains unknown. We exist within enormous sprawling, complicated ecosystems populated by thousands of interconnected species, but we don't fully understand how they interact or what role each plays. We can't know what really happens to that ecosystem if one of the species in it—an unremarkable-looking beetle, for instance, or a bat, or a frog, or an orchid—is removed forever. Most likely the ecosystem will continue to function. It will compensate and adapt in some unseen way. But how? And what does that tell us about the species that are left?

Most of the life on Earth, in other words, is still a mystery to us—undescribed and underexplored. Once it is named, a humble beetle doesn't perform its ecological role any differently. Nothing changes. But how can we hope to understand the complexity of life if we can't even identify its participants? How can we protect an animal we haven't named?

For centuries, natural history collections—and the intrepid collectors who built them—have helped us understand the biodiversity around us. Currently, taxonomists and biologists describe about eighteen thousand new species every year. Novel species are named every day. This includes extinct fossilized species and microscopic organisms like bacteria and viruses. First a holotype is identified—a single specimen that is used to describe and define the entire species. A new branch sprouts on the Tree of Life—an irreducible component of Earth's mostly still unknown biodiversity. Although eighteen thousand new species every year might sound like a lot, it represents an astonishingly small fraction of the world's total.

In fact, we are surrounded by unnamed species. At best, says Quentin Wheeler, a taxonomist and the president of the State University of New York College of Environmental Science and Forestry, we know perhaps one in four living insect species. We discover new species everywhere, from city backyards to remote rainforests and deep-sea environments. In September 2016 researchers at the University of Rochester in New York named *Lenomyrmex hoelldobleri*, a new species of tropical ant.[5] It was described from a single tiny specimen that had been flushed—already dead—from the stomach of a bright orange frog called the diablito, or little devil frog (*Oophaga sylvatica*). The ant lives in the Ecuadoran rainforest. Since there is only one sample and it came from a frog's stomach, no one knows precisely where it lives. Shark researcher David Ebert has named ten new species of shark from specimens he found for sale in just one Taiwanese fish market.[6] In thousands of other instances, though, biologists have already collected the novel specimens. The holotypes sit in collections, unidentified.

When you visit a large, established collection like the Field Museum of Natural History in Chicago or the California Academy of Natural Sciences—or museums in Leiden, or London, or Paris, or São Paulo—it is impossible to get a sense of the immense size and breadth of the collection that sits just beyond the public spaces. The bat collection at the American Museum of Natural History alone—just a small part of a much larger and more comprehensive mammal collection—includes more than 250,000 specimens. It continues to grow every year. Field biologists working in remote places capture bat specimens, then prepare them, label them with locality data, and accession them into the collection. Scientists from across the world travel to New York to measure and compare the specimens, like a vast reference library of biological material. Or they borrow specimens from the museum, the way Chatzimanolis borrowed beetles from the Natural History Museum in London, where coleopterist Max Barclay oversees an enormous beetle collection that encompasses about ten million individual beetles. More than a thousand new beetle species are described each year from that collection alone. In the United States, natural history collections contain an estimated one billion specimens. The entomology collection at the Bishop Museum in Honolulu, Hawaii, contains fourteen million specimens, including type specimens of thirty-six mosquito species. The Duke University herbarium includes more than 160,000 specimens of moss. The herpetology collection at the California Academy of Sciences in San Francisco has more than 300,000 cataloged specimens from 175 countries. At the Smithsonian Institution, the oldest botanical specimens date back to 1504.

The time that passes between the collection and description of a specimen is known as its shelf life or dwell time. According to a *Current Biology* article, the average shelf life across all orders of organisms is about twenty-one years.[7] Naming and describing species is a painstaking process that can take a long time.

Occasionally, though, the dwell time becomes much longer and turns into deep time. The specimens sit dormant for fifty or seventy-five years—sometimes a century or longer still. They wait in basements and storage cabinets. They sit in drawers or in jars of fixative. Their labels slowly turn yellow and fade. The person who collected them dies. But they still wait. Outside, on the street, where time still has its usual weight, the world changes. There are wars and there are scientific advances. The borders of the country where a specimen was collected are redrawn, then redrawn again, but in the collections the specimens remain unchanged. In most cases several million other specimens surround them, making them even more difficult to find.

A single specimen of an unknown species of spider might sit unidentified in a flask along with fifty examples of a known and commonly found species. In some small way, it is profoundly different from the others. It has something important to tell us about the ecosystem it came from and the processes that made it, but it has become almost impossible to find. It could remain undiscovered forever.

Then one day a graduate student who has become an expert in a particular genus of spiders takes the lid off the jar and sees something unexpected inside—something different.

Natural history collections are in danger, and many are struggling to survive.[8] In recent years a lot of institutions have seen drastic reductions in funding. The number of taxonomists working in collections has declined too. Much of the taxonomic work is now undertaken by evolutionary biologists, and they also are underfunded. The curators, vital custodians who oversee the care and organization of the collections, are disappearing as well. For instance, in 2001 the Field Museum had thirty-nine curators. Today there are twenty-one. The National Museum of Natural History in Washington, DC, has seen a loss of curators from a high of 122 in 1993 to a current low of 81: about a quarter. In the past, many larger institutions had a team of curators working in each discipline: three or four mammalogists, a couple of ichthyologists. There might have been several entomologists, each with a deep and informed interest in a single, highly specialized tribe of insects. Frequently these experts have been replaced by one overburdened collections manager. At some institutions, even collections managers are in short supply. The Field Museum currently lacks managers for several important collections: paleobotany, mycology, anthropology, and mammalogy.

Imagine a library filled with rare and important books but with no librarians to care for them.

Sometimes a natural history collection disappears altogether. At a smaller institution, or in an academic department, an expert in a particular taxon will retire or die and leave behind thousands of specimens. Occasionally those orphaned collections are adopted by larger institutions—enfolded into an established collection. But sometimes they're not.

With the advent of new technologies, many institutions have begun digitizing their collections, an ongoing process that will allow researchers worldwide to access specimens remotely.[9] Funded by the National Science Foundation, Integrated Digitized Biocollections— or iDigBio—is part of an ongoing global effort. The results are amazing. The iDigBio website includes links to more than seventy-two million records detailing individual specimens, housed in repositories across the world. There are collection data for everything from lichens and bryophytes to lowland gorillas. The Global Biodiversity Information Facility (GBIF) is another effort: a database funded by international governments that provides a single point of access to millions—currently more than half a billion—digitized records from institutions worldwide. Researchers at the Museum of Natural History in Berlin have begun digitizing its entomology collection, converting each specimen—via spherical high-resolution images—into a single incredibly detailed image that can be rotated and viewed from different angles. This means I can now sit on my couch and explore the gleaming surfaces of *Chrysis marqueti*, an iridescent green cuckoo wasp collected from the West Bank. Or, impaled on its pin in Berlin, *Hemidictya frondosa*—a butter-colored Brazilian cicada, frozen in midflight with its veined, leaflike wings outspread.

Overall, digitizing collections will increase taxonomic knowledge and help to reduce the hurdles that prevent species description. Known as the taxonomic impediment, the barriers to identification include a shortage of skilled taxonomists and curators, an insurmountable excess of undetermined material, a simple lack of interest, and constant reductions in institutional support. When the funding for projects like iDigBio stops—and it has been halted from time to time—the digitizing efforts halt too.

Either way, nothing can replace holding a specimen. The weight of the object in the hand engages the brain in ways that an image on a screen simply cannot.

For many people in the collections community, it feels as if natural history collections have been set adrift. People think natural history museums are quaint Victorian places, dimly lit and dusty—a throwback to an analog world. A cabinet filled with dead animals is no longer useful. After all, how can we learn anything from a 180-year-old beetle?

But the collections are more important now than ever. They provide an indelible taxonomic record of all life on Earth, and life on Earth is changing. According to many scientists, life is disappearing much faster than any time in recent history. By some estimates, current rates of extinction are a thousand times higher than natural background rates—much higher than previously calculated.[10] We have entered what some researchers have called the Sixth Extinction. For some species, the few specimens placed in natural history collections worldwide are the only examples that exist: living specimens are gone. Collections have multiple concrete uses. They allow researchers to understand the effects of complex processes like climate change, comparing the different species present in a particular region across time. We can use collections to pinpoint the introduction of invasive species and diseases to ecosystems. Ornithologists can track changes in bird migration patterns with collection data. Biological specimens—like the 250,000 bats at the American Museum of Natural History—might even allow epidemiologists to track the outbreak and cause of deadly zoonotic diseases like the ebola virus, which periodically erupt in human populations and then just as rapidly disappear into their animal reservoirs.

More important, the collections are repositories of human endeavor. They tell a profoundly human story. Three centuries of scientific thought can be condensed to a single unnamed longhorn beetle from Ghana or the long-vacated turriform shell of a Vietnamese land snail. The specimens come from across the world: from the bottom of the ocean; from inside volcanoes; from sub-Saharan Africa, the rainforests of Borneo, urban American backyards, and the endless brittle, frozen landscape of the Antarctic. A multitude of specimens. And there, in the collections, the mysteries await.

For a while they are lost—sometimes for centuries. But then, finally, they are found.

The Vertebrates

1

Pushed up a Mountain and into the Clouds: The Olinguito (*Bassaricyon neblina*)

Kristofer Helgen still remembers the moment he opened a drawer at the Field Museum of Natural History in Chicago, Illinois, and discovered a new species inside.[1] It was December 2003, bright and cold outside—a northern winter. The crowded city streets had been scoured clean by winds gusting toward the city across Lake Michigan.

In the Field's extensive mammal collection, it was warm and quiet. There is something hermetic about the collection, deep within the museum's interior, insulated by thick walls—a warren of brick-bounded rooms with no windows. A few other researchers were working quietly at desks, carefully measuring a row of identical-looking brown bats or hefting a walrus skull from a cabinet as if it were a bowling ball. A researcher at the University of Adelaide Environment Institute, Helgen stood before a cabinet stretching from floor to ceiling. Several drawers supposedly contained the skulls and preserved skins of a small, arboreal raccoon-like mammal called the olingo, which lives in the remote cloud forests of Central and South America. Helgen had traveled to the Field Museum specifically to examine its olingo specimens, but when he pulled open the drawer he knew he'd found something else instead.

"There were these gorgeous soft, thick red-furred things," he tells me. "I knew instantly they weren't olingos. They're nothing like anything anyone has ever described or put a name on."

Eventually, after a decade of careful investigation, which included a field expedition to northern Ecuador to find the animal in the wild, Helgen named the new species the olinguito. In a formal description published in 2013, he gave the species the scientific name *Bassaricyon neblina* from *neblina*, Spanish for mist. He found it in the mist. He named it, and he made it real.

"It's a perfect illustration of how something can hide in plain sight when you just don't know what you're looking at," says Helgen.

If anyone in the world was qualified to discover the olinguito—undocumented in an otherwise exhaustively well-cataloged museum collection like the Field's—it was Helgen. In the past decade he has named more than thirty new mammal species and subspecies, all from archived specimens he and his collaborators have found in museum collections.

In fact, he has been preparing himself for this research for more than half of his life. "I became fascinated in my earliest childhood with the question of how many kinds of mammals there are in the world," says Helgen. "By the time I was ten years old or so I knew the scientific name of almost every mammal."

At thirty-seven, Helgen is still young, with a mop of red hair not unlike one of the olinguito pelts at the Field Museum. His ability to detect subtleties between specimens is well known. When he was still a child, Helgen says, he had already identified several taxonomic blind spots—entire genera that remained unresolved. No one knew for certain how many species they even contained. "At the top of that list," he says, "was *Bassaricyon*, the genus of the olingos."

The specimens: when a museum specimen is collected in the field, it is skinned and prepared. It begins the process of becoming a zoological artifact that can retain its meaning for centuries. The soft tissue is removed. The bones and skull are placed in a case filled with dermestid beetles, which efficiently strip the flesh in a matter of hours and leave behind a meticulously cleaned skeleton. In the current era of molecular biology, when a specimen is collected in the field researchers might take multiple tissue samples and freeze them for later molecular analysis of its DNA. But when the first olinguito specimens were collected in Ecuador and Colombia, no one could have foreseen the advent of molecular technology. No one even knew what DNA was—research on its structure was still decades away. Back then the specimen just became a pelt in a drawer with a skull and a disarticulated skeleton in a box beside it. Each bone has the specimen number carefully inscribed on it by hand. In their drawer, the olinguito skins look like a collection of bright red stoles laid side by side. Bird skins and mammals are stuffed with wads of white cotton in the eye sockets. Insects are mounted on pins. Fish and amphibians are either stored in liquid preservative or dried.

Sometimes there is no skin at all: only a skull has survived, or just part of the skull—a broken fragment and maybe a few bones. A beetle specimen could be missing its abdomen; a gecko might have shed its tail and left a ragged stump; the frayed wing of a bat specimen is marked by a perfectly round pellet hole—a reminder that a field collector shot it from the sky, watched it tumble through the air, and collected it where it fell. A researcher must work with whatever material is in the collection.

The *Bassaricyon* genus had remained a puzzle since it was first erected in 1876. Helgen decided to solve it, first methodically identifying its members, then gradually filling in what he believed were the remaining gaps in the record. In the past, he says, some species had been named and quickly forgotten, not properly incorporated into the family tree; others were what he now calls oversplit—when a single species is incorrectly divided into several species that don't actually exist. To makes sense of it, Helgen knew he would have to search museum collections worldwide, locating, comparing, and measuring as many olingo specimens as he could find.

"I learned to tell the olingos apart," he says. When he opened the drawer at the Field, expecting to find it filled with olingos, Helgen says his first thoughts were, "What am I missing? Is this the wrong place? What is this creature?"

He scrutinized the specimens, absorbing their smallest details.

"I opened the skull boxes that were next to them and, sure enough, it is a procyonid, and it is very much like an olingo," he says. "But something completely different."

And even I, untrained in detecting the subtle—often almost imperceptible—morphological differences between very similar species, can appreciate that olingo and olinguito skulls are not the same.

The differences are narrow but deep. Olinguitos are smaller, weighing about two pounds, the smallest members of the raccoon family. They have smaller teeth and numerous subtle differences in their skull morphology. They live at higher elevations than the olingos, between five and nine thousand feet above sea level. Their pelts are longer and denser than those of their closest relatives—a specific adaptation to their habitat high in the elevated rainforests of the northern Andes.

"You might expect it to live on a single mountaintop," says Helgen. "Maybe that explains why everybody had missed it. But this thing is widespread in the northern Andes. And then you might expect that there's not a lot going on with variation within it, but it turns out there were four very different kinds of olinguito. We were able to name them all as subspecies."

In fact, specimens of all four of the newly named subspecies have spent decades in collections, unnoticed until now. The holotype of the nominal species—its full scientific name is *Bassaricyon neblina neblina*—resides at the American Museum of Natural History in New York: specimen M-66753. As with all holotypes, regardless of institution, it is now clearly marked with a red tag to denote its primacy as a type specimen, and securely locked away. A holotype is irreplaceable and is subject to the highest security. The specimen was collected at Las Maquinas in northern Ecuador on September 21, 1923, by George Henry Hamilton Tate, an American field biologist. It spent ninety years stored in the collection. The first olinguito was collected even earlier, in June 1898. That specimen is part of the American Museum of Natural History's mammal collection too. Examples of the three other olinguito subspecies are part of the mammal collection at the Field Museum—all collected in Colombia in the 1950s. Suddenly the entire *Bassaricyon* genus had to be revised to incorporate its newest members.

There is a photograph in Helgen's description of the olinguito, taken on September 6, 1951—a black-and-white image with a strange greenish tint. In it an Ecuadorian hunter, bearded and barefoot, wearing patched jeans, squats on the grass near the edge of the jungle in San Agustín, Colombia. He squints up into the camera. In each hand he holds a dead olinguito. Tied to a string across his chest is a long-tailed weasel (*Mustela frenata*). The photo was taken by Philip Hershkovitz, a prolific curator and specimen collector who spent fifty years at the Field Museum until his death in 1997. The body of each olinguito is about the size of a house cat, ending in a straight brushlike tail that is longer than the rest of its slender, sinuous body. One of the animals the hunter holds became—sixty-two years later—the holotype for another subspecies, named *Bassaricyon neblina hershkovitzi*. In other words, the photograph captures the exact moment a new species was discovered. It also marks the moment the discovery was overlooked—the precise moment it slipped away.

It had been overlooked before, and it would be again. In the 1960s and 1970s, says Helgen, an olinguito was part of several zoo collections. It made different vocalizations than other *Bassaricyon*. It refused to breed with its olingo cagemates, so it was moved endlessly from zoo to zoo—from St. Louis to Washington, and on.

But it was a misidentified olinguito.

The four olinguito subspecies are distinct morphologically—Helgen's hundreds of precise museum measurements prove that. But they live in different places, too. Eventually it became clear that the olinguitos each occupy their own subtle ranges and territories. One lives in the forest canopy in northern Colombia, on the western slopes of the Western Andes; another is found to the south and east, across the desolate high-altitude passes of the Western and Central Andes mountain ranges, on the eastern slopes of the Central Andes; the third subspecies lives southward, in the mountainous parts of northern Ecuador. Finally, another more mysterious subspecies seems to inhabit the gaps left by the other subspecies, living in the porous, poorly defined spaces between them.

Helgen formulated their distinct territories by plotting the collection data of each museum specimen on a map. "I drilled down and made sure I understood the complex northern Andean geography," he says. "I looked it up in maps and gazetteers and things. The situation crystallized, and I realized these different slopes and different sections of the Andes were correlated with these different morphologies."

In August 2006 Helgen assembled a small team of zoologists and traveled to Ecuador to try to find the olinguito in the wild. Roland Kays, director of the biodiversity laboratory at the North Carolina Museum of Natural Sciences, accompanied Helgen to the Otonga Nature Reserve on the western slopes of the Andes in northern Ecuador. "They're very steep mountains, and they're covered in this cloud forest," says Kays. "The higher up you get, the more dwarfed the trees get, but there's still some fairly big trees. It's like the tropical rainforest, but pushed up a mountain and into the clouds."

A few years before, he says, during fieldwork for his graduate studies, he'd published some early data on the olingos. In the process, he had become the de facto world expert on the *Bassaricyon* genus. Until then no one had studied them closely at all. For three weeks in 2006, on steep slopes covered with fig trees, Kays and Helgen conducted an extensive mammal survey. They erected mist nets for bats and set traps in the understory for ground-dwelling mammals. And they both craned their necks at night peering into the dimness, shining flashlights into the mist-choked canopy to search for olinguitos.

"There's all sorts of crazy, diverse species," Kays says. "It's hyperdiverse. You get different highland versions of what's in the lowlands. That's what the olinguito is, basically: it's a mountain-specialized form, because it's so cold up there. It's not freezing cold—it's not ice cold—but it's not the tropical weather you think of for the rainforest, and it's very damp."

The damp Ecuadorian fig tree forest: a strangely canted world, everything growing skyward from the green slopes, as if from the deck of a sinking ship. Helgen sighted the olinguito in the trees almost the moment he arrived. "We found it on the first night we went down to look for it," he says.

There are all sorts of reasons it has managed to stay hidden for so long among the clouds, in the tangled fig trees of Otonga Reserve. "It only comes out at night," he says. "It doesn't really come out of the trees. It's pretty shy, and it looks to the untrained eye either like an olingo or like a kinkajou—and those two other types of animals live in the same geography."

Like the olingo and the kinkajou, the olinguito has adapted perfectly to its environment—a life spent in the canopy of the cloud forest. "It has a long tail," says Kays. "It's a very good tree climber. It has very good balance, and it's still light enough that it can jump from branch to branch. And it has these big eyes because it's doing this high-wire act at night when there's absolutely no light. Imagine not just walking around the forest floor at night but jumping from tree to tree."

Occasionally taxonomists discover what they call a cryptic species: a species that is genetically distinct but so similar morphologically to an existing species that the two are indistinguishable. With the advent of DNA sequencing technology to detect subtle genomic differences, the discovery of cryptic species has become much more frequent. In September 2016 researchers at Goethe University in Frankfurt, Germany, announced that the giraffe—one of the most morphologically distinct species in the world—is actually a wide-ranging group of four different species.[2] They don't interbreed at all. In fact, the four species are as genetically distinct as the polar bear (*Ursus maritimus*) is from the brown bear (*Ursus arctos*). But the olinguito is not a cryptic species. The moment Helgen opened the drawer at the Field Museum, he knew he'd discovered a new species. It diverged from its closest relatives an estimated 3.5 million years ago, evolving its own distinctive characteristics. Helgen's discovery made headlines across the world. The olinguito was the first new species of carnivorous mammal named in the Americas for thirty-five years.[3]

"What is it when we add one more species?" Helgen asks. "For mammals, there are only about six thousand of them. There are a huge number of entities and agencies and people on this planet who take great interest in knowing something about each of those six thousand mammal species. Every time I name one of these species it enters the pipeline and people start to think more about it—try to learn more about it. It gets on endangered species lists at the national and provincial level."

Helgen and Kays estimate that about half of the olinguito's range has already been deforested—urbanized, denuded, or converted to farmland. "These ecosystems that are out there," says Helgen, "we understand them extremely poorly. They're unbelievably complex, much more so than anything man-made. To really get how they work, we need more than the embarrassing misunderstanding of not even seeing the olinguito that's already there—right in front of us."

2

Beneath a Color 83 Sky:
The Ucucha Mouse
(*Thomasomys ucucha*)

Essentially, *Thomasomys ucucha* looks like a mouse from central casting. Its coat is dark brown, gradually becoming light gray toward its soft underbelly. Its body is small and nondescript. Even Robert Voss, the American Museum of Natural History's curator of mammals, who named and described the species in 2003, calls it "a boring-looking brownish, greenish rat-shaped mouse."[1]

Voss first saw *T. ucucha* in northeastern Ecuador in 1978 as a graduate student at the University of Michigan. When he examined it—in the high cold air of the Cordillera Oriental—he says he knew immediately that it was unknown from its prominent and distinctive teeth.

When describing species, field biologists like Voss often use a color catalog to accurately describe a specimen's external coloration, comparing what they see in real life with color chips on a page. The standard text is Frank Smithe's *Naturalist's Color Guide*, published in 1975. Using Smithe as his guide, Voss described the mouse thus: brownish olive (color 29) on its flanks, turning dark neutral gray (color 83) on its belly, with a superficial wash of light neutral gray (color 85) or glaucous (color 80). The enamel of its long, prominent incisor teeth is marked with bands of spectrum orange (color 17).

Voss was in Ecuador to collect specimens of semiaquatic insectivorous mice for his dissertation research. It was, he says, a challenging environment. At dawn the banks of the tributaries that tumbled downhill were frozen, and there was rime on the grass tussocks. Sudden blizzards would advance across the landscape without warning, obscuring the hillsides and killing the sturdy cattle the local herders grazed there. Even on bright, clear days, the steep slopes quickly turned cold the moment the sun passed overhead on its arc.

In his formal description of the mouse from 2003, Voss writes: "Steep, unstable slopes afford few places to camp, and the dripping-wet forests are miserable places to run traplines for more than a few days at a time." The conditions help explain why so little is known about the biodiversity of the region. Put simply, no one wanted to be there. It was cold and wet—an extremely difficult place to work. Imagine for a moment trying to work for more than a day or two tilted at a forty-five-degree angle. As Voss put it in his paper, "Most collecting expeditions to eastern Ecuador have hurried downslope to more inviting habitats in the Amazonian lowlands."

Instead, Voss remained upslope, on the high roof of the world, beneath a color 83 sky.

At about fourteen thousand feet, he began to find what he calls "a big, buck-toothed species" of rodent: *Thomasomys ucucha*. Voss set commercial rattraps near the treeline of the stunted subalpine rainforest, in a wet, mossy tangle of low branches called elfin forest. He placed other traps along the ice-encrusted riverbanks. Over several days, he collected forty-three *T. ucucha* specimens and brought them home. Then the specimens waited. They sat undescribed at the University of Michigan Museum of Zoology in Ann Arbor until 2003. The species holotype—an adult male specimen trapped on April 26, 1980, in the valley of the Río Papallacta—was among them.

But that was 1980. In the intervening years Voss had busied himself with other projects. In 1985 he began working at the American

Museum of Natural History in New York. His primary research interest is the evolution of marsupials. In all, he has participated in naming eleven other mammal species. Describing a new species is not a simple process, he says. In fact, it can be painstakingly slow and arduous, requiring morphological comparisons with numerous specimens from closely related species. The uniqueness of a new species must be empirically confirmed beyond all doubt.

So *T. ucucha* waited its turn in a drawer in Michigan. Perhaps some people find the passage of twenty-five years between the collection and description of a new species excessive. It's not; in fact it's average. But *T. ucucha* had waited longer than that—much longer. When Voss finally turned his attention to the unnamed specimens he'd collected in Ecuador in the 1970s, he arranged to borrow them from the University of Michigan—a common practice in taxonomic studies. In the meantime, he searched the enormous and encyclopedic collection of mammal specimens at the American Museum of Natural History, pulling examples of a few closely related species to compare with his specimens arriving from Michigan. He found the specimens he needed, but he found more than he expected: he found his mouse. There were three of them: brownish olive flanks and gray underbellies, their teeth marked with bands of spectrum orange.

They had been collected in 1903 in Ecuador by Ludovic Soderstrom, an amateur field biologist—or more likely, says Voss, by one of the many Ecuadorian natives Soderstrom had trained to collect specimens for him. For more than four decades, until his death in 1927, Soderstrom employed a small collecting army, which dispersed across the steep terrain gathering up every living thing it encountered. A foreign diplomat, Soderstrom worked for the British Consulate in Quito, and he owned large country estates in the eastern and western lowlands of Peru. Prolifically supplied by his collectors, Soderstrom sold and donated bird and mammal specimens to museum collections across the world.

The list is encyclopedic: he sent specimens of Ecuadorian mammals to the British Museum in 1896; a large collection of hummingbird skins to the Perthshire Natural History Museum in 1900; passion-flower seeds to the United States Department of Agriculture; a silky shrew opossum (*Caenolestes fuliginosus*) to the American Museum of Natural History; and a yellow-collared tanager (*Tangara pulcherrima*) to the Harvard University Museum of Comparative Zoology in 1914. He sent live samples of an Ecuadorian cactus, later named *Opuntia soederstromiana*, to the Carnegie Institution for Science, but they all died in transit. He sent neotropical deer specimens to the Field Museum, seeds to Kew Gardens in London, and endlessly on and on.

In 1911, when American ornithologist Samuel Rhoads visited Soderstrom in Quito, he found the sixty-eight-year-old in his extensive bachelor quarters in the middle of the city. In the retelling, it sounds like a home from a dark and strange story. From the house Soderstrom had a view of the foothills of the volcano Mount Pichincha, which rose into the air four miles away. He had built two botanical gardens and filled them with specimens from his expeditions. Orchids grew everywhere.

"To our great surprise and pleasure," wrote Rhoads, "here also the humming birds were at home, feeding and playing and even building their nests in the trees and against the walls of the buildings, as many as ten or twelve species having favored the old gentleman with their visits and four or five species making his home their own also. In the house also resides a beautiful indigenous long-tailed mouse with grooved front teeth which was named after Soderstrom by Prof. Thomas of the British Museum. One of these he caught for my collection and presented it to me with his compliments; a very great favor, I can assure you."[2]

In 1921 Soderstrom gave the American Museum of Natural History curator Harold Elmer Anthony a large collection of specimens: more than fifteen hundred mammals and two thousand birds. Among the specimens were three nondescript mice trapped in the highlands near Papallacta: two males and a female. The other details are lost. At the time the mice were identified only to genus— *Thomasomys*. These are the three forgotten specimens Voss found archived in the American Museum of Natural History's mammal collection in 2003. Named for the British zoologist Michael Rogers Oldfield Thomas— who described more than two thousand mammal species and subspecies — the *Thomasomys* genus is a diverse group of rodents. In total it includes almost forty species: the beady-eyed Oldfield mouse, the ashy-bellied Oldfield mouse, the strong-tailed Old-field mouse, and the wandering, snow-footed, and distinguished Oldfield mice are among them. In time Anthony named several *Thomasomys* species himself, but he never revisited the unidentified mouse he obtained from Soderstrom in 1921 in Quito. When Voss finally described the mouse in 2003, he gave it the species name *ucucha* — the local Quichua word for mouse. By then the *T. ucucha* specimens had been part of the American Museum of Natural History mammal collection for almost a century. They were already there in 1978, lined up neatly in a drawer, when Voss was searching the wet Andean slopes in the rain, looking in vain for a comfortable and level place to make camp near the edge of the elfin forest.

T. ucucha is not a charismatic mammal species like a snow leopard or a koala. It's not an apex predator either. It's not dominant or powerful. It's a humble mouse, a lowly species: a boring-looking brownish, greenish rat-shaped mouse that lives on a remote crest of the northern Andes, at the tangled edge of the mossy subalpine forest. On the slopes it serves mostly as prey for the hawks and weasels and quietly eats its share of worms and insects. Does this lessen its importance somehow? Not on the mountainside. Up there *T. ucucha* quietly fulfills its role, even though Voss concedes he's not fully certain what that role is.

"We still don't know very much about it except that it apparently has a very small geographic range in a very poorly known area," he says. Regardless, it is unique — profoundly different from the other thirty-two species of small mammals Voss found sharing its habitat on the rugged Cordillera Oriental near Papallacta.

Or, as Voss grudgingly puts it, sounding underwhelmed by his discovery: "Every species has a unique place in the polity of nature."

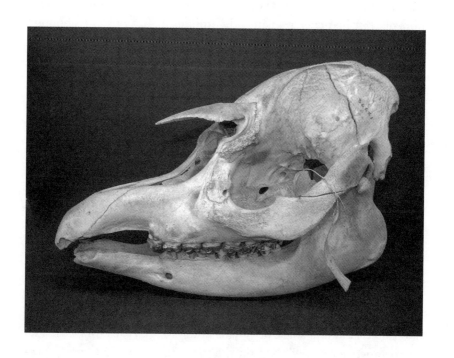

3

Going on a Tapir Hunt:
The Little Black Tapir
(*Tapirus kabomani*)

Mario Cozzuol is standing in the green gloom of a rainforest clearing in Rondônia State in northwestern Brazil, swatting mosquitoes away from his face. He crouches low, searching a well-used trail for fresh dung. Moving slowly and quietly, he crosses the clearing. The forest rises in front of him like a wall. Cozzuol wipes his bald head and scours the ground for freshly broken tree stumps, but there is nothing. He swats at the humid air again and keeps moving. Cozzuol is searching for a living specimen of *Tapirus kabomani*, a novel species of tapir he described in 2013.[1]

The process of naming *T. kabomani* began more than a decade earlier with a study of tapir fossils, says Cozzuol, a paleontologist at the Universidade Federal de Minas Gerais, in Belo Horizonte, Brazil. Working alongside a student, he requested loans of tapir skulls from scientific collections across Brazil. He needed the skulls of living tapir species to compare with skulls of fossilized tapir remains excavated from sites across the Amazon. When the skulls arrived, the student lined them up on a counter and began to inspect each one. "In this process," says Cozzuol, "she comes to me with a skull and says, 'This specimen is weird. I don't know what it is.'"

That single skull was different from the others, he says, in many subtle morphological characteristics. It wasn't a lowland tapir, or *Tapirus terrestris*—a relatively common and widespread species. At first Cozzuol wondered if it belonged to a species of mountain tapir instead. "When we recovered the story of this animal, this particular specimen, we found that it was hunted in a river settlement across from the Porto Velho town in Rondônia State, in the Amazon."

The tapir is a stout, large-bodied herbivorous mammal with a range that extends across South America. It is the largest land mammal on the continent. With its sturdy build, it looks a lot like a pig with a long, mobile snout. Sometimes it is hunted for food by the indigenous people who live in the rainforest and by the ribereños, the non-Amerindians who inhabit the riverbanks. The specimen that yielded the unusual skull had been killed, Cozzuol says, by riverine people from a settlement near Igarapé Belmont, about eight miles north of downtown Porto Velho, the capital of Rondônia State. A city with about 500,000 residents, Porto Velho sits right on the edge of the rainforest—encroached upon by the vines. After they killed the tapir, most likely they butchered it there and carried away slabs of meat to be eaten or salted and sun-dried. The tapir's remains were found near the river, with a bullet hole in the braincase. No soft tissues were left. The skull is now part of the mammal collection at the Universidad Federal de Rondônia.

The location where the tapir was killed is a lowland region, suggesting the specimen wasn't a mountain tapir, or *Tapirus pinchaque*— a much less common tapir species. "This cannot be a mountain tapir in the lowland Amazon," explains Cozzuol. "We checked it more and we saw differences with mountain tapirs. We arrived at the conclusion that this was probably a new species. But the only thing we had was just one specimen, and this was obviously not enough— and just its skull."

Cozzuol began to measure and carefully compare the morphology of the other tapir species. In all, he says, he examined more than seventy skulls. Eventually he described *T. kabomani*—the little black tapir.

The indigenous people who live in the settlements around Porto Velho are matter-of-fact about the existence of the new tapir, says Cozzuol. They know it exists and can identify it quickly in the wild. To them it differs from the common lowland tapir in obvious ways: first, it's shorter and smaller, with a distinctive black mane and a broad forehead. With an estimated body weight of about 250 pounds, it's only about one-third the size of the lowland tapir. To find another specimen of the new species, Cozzuol decided to work with the indigenous people. In December 2009 he obtained a license to collect from the field and hired a local hunter to kill a specimen for him. The hunter pointed to the north and told Cozzuol, "We have to go across the river. If you want, I can get you there."

There, in southern Amazonas State a few miles north of the Rio Madeira, which snakes and elbows its way northeast across the continuous rainforest, the hunter killed the specimen that eventually became the holotype of *T. kabomani*. "We collected the skull," says Cozzuol. "The skull matches the new species."

But there is another skull from the new species, says Cozzuol—it's much older than the holotype. It's been part of the mammal collection at the American Museum of Natural History since 1914. Theodore Roosevelt collected the skull, along with a partial skin of the animal it belonged to, during an expedition to the Amazon. In December 1913 the twenty-sixth president of the United States traveled to the interior of Brazil. His goal was to locate the headwaters of the Rio da Duvida, the River of Doubt—a dangerous and uncharted tributary of the Amazon. Roosevelt was then fifty years old. The second term of his presidency had ended in 1909; in 1912, he'd mounted an unsuccessful campaign for a third term. A bullet from a failed assassination attempt a year earlier was still lodged in his chest—doctors deemed it too dangerous to remove. Accompanied by his son Kermit and the famous Brazilian explorer Colonel Cândido Rondon, in January 1914 Roosevelt undertook a perilous journey on foot into unknown parts of the Amazon. They traveled by dugout canoe along tea-colored Amazon tributaries, the jungle pressing in from all sides. With them they took two naturalists from the American Museum of Natural History—an ornithologist and a mammalogist.

"On the morning of January 9th we started out for a tapir-hunt," Roosevelt wrote in *Through the Brazilian Wilderness.* "Tapirs are hunted with canoes, as they dwell in thick jungle and take to the water when hounds follow them."[2]

The expedition party had left Cáceres, a settlement near the Bolivian border, heading north up the Rio Sepotuba, the River of Tapirs, a swift, clear stream that descended from the high country into lowland jungle. Roosevelt wrote: "The tropical forest came down almost like a wall, the tall trees laced together with vines, and the spaces between their trunks filled with a low, dense jungle."

Suddenly, north of Porto Campo, there was activity on the river. The hounds had sent a large male tapir crashing into the swift water. "The tapir was coming down-stream at a great rate, only its queer head above water, while the dugouts were closing rapidly on it, the paddlers uttering loud cries. As the tapir turned slightly to one side or the other the long, slightly upturned snout and the strongly pronounced arch of the crest along the head and upper neck gave it a marked and unusual aspect."

Roosevelt shot it in the body, but the tapir escaped and the hounds followed. The wounded tapir ran into thick jungle and made its way upstream, where it promptly turned and fell into the water again. "I shot it, the bullet going into its brain, while it was thirty or forty yards from shore," Roosevelt wrote. "It sank at once."

After the body was collected from the river—it was a lowland tapir—the specimen was sent to the American Museum of Natural History. But it wasn't the only tapir Roosevelt obtained during the Amazon expedition.

"The remaining members of the party killed two or three more tapirs," he wrote. "One was a bull, full grown but very much smaller than the animal I had killed. The hunters said that this was a distinct kind. The skull and skin were sent back with the other specimens to the American Museum, where after due examination and comparison its specific identify will be established."

A century after it was killed, Cozzuol has inspected the skull of the smaller bull tapir Roosevelt mentioned in his account of the expedition-the distinct kind. Specimen AMNH 36661, it has remained at the American Museum since then. Although badly damaged, Cozzuol says, the skull shares important cranial features with the holotype the hunter killed in southern Amazonas State in 2009.

It's a *T. kabomani*. "It's very interesting the good eye Roosevelt had for these things. He was a hunter—a very good naturalist for the time. He noted the differences."

In all, Roosevelt and his party collected about 450 specimens of rainforest mammals, representing almost a hundred different species. They spanned a wide array of Amazonian mammal diversity: nine-banded armadillos, capybaras, jaguars, cottontail rabbits, white-lipped peccaries, and marsh deer. In March 1914, on their arrival at the American Museum of Natural History, the specimens were assessed by Joel Asaph Allen, curator of mammals. Ultimately, when Allen inspected the tapir specimens, he determined that the skulls were all from common lowland tapirs.[3] Cozzuol sent me a photograph of the 1914 skull. The rear part is broken off and sits by itself like a strange bowl. The specimen number—36661—is handwritten on the skull in pencil, faded but still visible near the parietal suture. The partial skin is folded into a rectangle like a misshapen floor rug, pale and leathery. Along one edge of the folded skin, the shape of the tapir's head is clearly visible—mouth open, long trunklike nose extended upward. A dark mane runs from its forehead along its head and shoulders like a low crest. To Cozzuol, it is distinctive.

Whenever possible, modern taxonomists and field biologists take tissue samples from specimens. Later, in a laboratory, they can extract DNA and sequence its genome. Cozzuol did this. Comparing the DNA sequences of certain genes helps researchers determine the phylogenetic, or evolutionary, relations between different organisms. Two species might look different from one another, but it's important for taxonomists to understand just how different they are genetically and how they're related. The molecular approach to species identification has given rise to a technique known as DNA

bar coding. Researchers extract DNA from a validated type specimen and amplify it, sequencing a fragment of a mitochondrial gene
called cytochrome *c* oxidase 1, or CO1.

Worldwide, scientists have embarked on the International Bar-
code of Life (IBoL) project—an initiative that includes researchers
in twenty-five countries. The long-term goal of the project is to generate a species-defining genetic bar code for every known species
on Earth and store it in an ever-expanding database. Each bar code,
like the bar code on a product in a grocery store, is a unique identi-
fier. In the near future, a researcher will be able to bar code a series
of unidentified specimens and determine that this specimen belongs
to species X, that is species Y, this is species Z, and so on. This approach is much faster than traditional taxonomy, but it has its limi-
tations. If a type specimen exists for a particular species, generating
a bar code for a specimen might be useful: it would allow for immediate identification. But what happens if the DNA from a specimen collected in the field has no match? Is it a new species? How
different does a specimen's DNA need to be to constitute naming it
as a novel species? How is it different? Which species came first, and
what is its relation to its closest relatives? DNA bar coding answers
none of these questions, and they're important.[4]

Describing a new species is not a straightforward process. History
is littered with organisms that have been overdifferentiated—split
to produce species that don't really exist. Later those species that
are really not species have to be abandoned. In other cases species
haven't been differentiated enough: an unnamed organism is hidden within the name of another species that it closely resembles.
For instance, in 2014 Field Museum mammalogist Bruce Patterson
discovered that the little yellow-shouldered bat (*Sturnira lilium*), a
common and widespread species found across South America, is
actually a constellation of seven species—distinct genetically but
very similar in appearance. Almost none of the researchers who
studied the little yellow-shouldered bat, and there are many, had
actually seen one. And there were new species to properly characterize. When taxonomists name a species, they are attempting to

classify and organize the natural world. When they get it wrong, their errors can persist for centuries.

Cozzuol is convinced that his description of the little black tapir is valid, but not everyone agrees. Even now, several years after the description was first published, he seems unprepared for the level of criticism he received from his peers. Among the dissenters are Robert Voss,[5] who described the ucucha mouse in 2003, and Kristofer Helgen, who described the olinguito in 2013.[6] Voss suspects that the small dark tapirs are simply subadult specimens of the larger and more common lowland tapir. Different, but only temporarily. It's difficult to know for sure, he says, because there are no voucher specimens for the novel species—another important criticism. A voucher specimen is a representative specimen that is deposited in a collection and used as a reference. It can be examined by anyone who wants to validate or refute a researcher's claims. In this case, says Voss, while there are several supposed voucher specimens, Cozzuol never sequenced the DNA of the same animals he used to define the morphological traits of the suspected new species.

The lowland tapir occupies low-altitude regions of South America from Colombia and Venezuela in the north southward all the way to northern Argentina. In other words, it has an enormous geographic range, filled with many diverse and different habitats. "Species with very large geographic ranges often show geographic variation," says Voss.

The degree of genetic variation Cozzuol had seen between specimens that belong to his candidate species and the lowland tapir specimens he sampled is not unusual, says Voss. It's the sort of variation that typically can be found between individuals of a single species. Importantly, despite its extensive range, a single tapir doesn't travel far at all. It's a slow-moving and cautious animal. This sort of situation eventually gives rise to different species. A lowland tapir collected from Venezuela, in the northern limits of its range, might differ significantly from an Argentine tapir collected more than two thousand miles to the south—a very different habitat. This might be true, but then again, it might not: very little is known at present about the genetic diversity of tapirs. The uncertainty worries Voss, who suspects that the mountain tapir—with its distinctive thick, woolly coat—might also be not a distinct species but just a highland variant of the lowland tapir.

Voss is not alone. The gaps in tapir taxonomy unsettle Cozzuol too.[7] "In doing the genetic work," he says, "we found that what we now call *Tapirus terrestris*, even after removing *Tapirus kabomani* from it, is probably not a single species but a complex of species, or at least a very diverse species with many different populations inside it."

Species descriptions, and the intricate many-branched phylogenetic trees that often accompany them, are provisional—subject to change. A species is nothing more than a hypothesis. Approached this way, Cozzuol's hypothesis is simple enough: the little black tapir skulls he has collected in the Amazon are not the same as those of the common lowland tapir. They're different. While subtle, the morphological and genetic differences are sufficient—and consistent enough—that the two animals can be considered different species. They represent separate lineages that don't interbreed. The genetic data, says Cozzuol, suggest that the little black tapir diverged from the lowland tapir recently—about half a million years ago.

The local people have no problem separating the little black tapir from the more common lowland species. But indigenous people—like the riverine tribes that live on the banks of the Rio Madeira and Igarapé Belmont—have a habit of overdifferentiating game species, says Voss. More precisely, in their green world they differentiate species in ways that are meaningful to them. As hunters, their classification system is based on distinctions that allow them to predict the tapir's behavior. The tribespeople are not concerned with our approach to taxonomy and delimiting species—and they shouldn't be. A tribe in Peru recognizes at least seven species of tapir, all of them probably the lowland tapir. Unswayed, Cozzuol says many new species have been described in the past few decades with less supporting evidence than he has provided. In several instances, species have been described from a single specimen. Even Voss has done this, naming *Monodelphis sanctaerosae*, a new species of short-tailed opossum, in 2012 from a single representative.[8] In 2006, Helgen named a novel species of Samoan bat from a single cracked skull that had been collected in 1856 and was part of the neglected collections at the Academy of Natural Sciences of Drexel University in Philadelphia.

What if there is no specimen at all? In 2005, when primatologist Trevor Jones named the highland mangabey monkey (*Lophocebus kipunji*) he did so using a photograph of the holotype, taken near Mount Rungwe in Tanzania.[9] There were no voucher specimens to assess or measure. There was no skull. With no tissue samples, there was no DNA to extract either. In the descriptive paper in *Science* the authors noted that *L. kipunji* looks different from closely related species and makes a distinctive "honk-bark" call instead of the familiar "whoop-gobble" made by other species in the genus. But is it really a species? Is it just a subspecies? Is it a geographic variant? Does it just belong to a small fragmented group of monkeys that, for some reason, have decided to "honk-bark" to one another? These distinctions, between a species and a subspecies, or a local variant, are incredibly important. They get to the heart of taxonomy and classification. And they help to determine how scientists from numerous different disciplines view the animal—from taxonomists and evolutionary biologists to systematists, ecologists, and conservationists.[10]

A year later, Tim Davenport of the Wildlife Conservation Society refined the taxonomy of the species further. He renamed the monkey *Rungwecebus kipunji*, erecting an entirely new genus for it— the first new primate genus in eighty-three years. Davenport examined the single known museum specimen—a male subadult that had been found dead in a farmer's trap in a maize field near Mount Rungwe and is now part of the collection at the Field Museum in Chicago. Subsequent DNA analysis proves it is distinct from any other species. But the holotype remains a monkey in a photograph: an ash-colored specimen crouched on a moss-covered log with its long tail curled in the air like a question mark.

Is that good enough?

Ultimately, a new species is accepted by consensus. There is no central governing body that affirms newly described species. Taxonomists either accept a candidate species and incorporate it into their work or they don't. Founded in 1895 and based at the Natural History Museum in London, the International Commission on Zoological Nomenclature (ICZN) has established a system to provide uniformity to the process of naming new species. Taxonomists refer to it simply as the Code. But the Code can often be ambiguous and opaque—byzantine in its articles, subclauses, and amendments. As with the United States Constitution, two researchers might read its language closely but come to opposite conclusions about its exact meaning. And so, even with the Code in hand, taxonomists have failed to reach a consensus on the little black tapir.

To further complicate the issue, Marc van Roosmalen, an independent Dutch-born primatologist living in Manaus, in Amazonas State, claims he discovered the tapir first. He tentatively named it *Tapirus pygmaeus* a few months before Cozzuol gave it his name. Van Roosmalen says he has been collecting data on the tapir for more than a decade. His description was rejected by reviewers who suspected—as they have with Cozzuol's work—that he was describing a subadult lowland tapir instead. For his holotype, van Roosmalen described a female adult, shot and eaten by locals from a village called Tucunaré, on the Rio Aripuanã.

Van Roosmalen, who lives in the Amazon, keeps the skull hidden, prevented from depositing it in a collection by a complicated relationship with Brazilian lawmakers. In 2000 *Time* magazine named him one of its Heroes of the Planet. But in 2002 he was charged with bioterrorism by Brazilian authorities.[11] Allegedly he had violated Brazil's restrictive rules governing the collection of scientific material from the rainforest. Eventually the charges were dropped, but later he was charged again, in a federal court, this time with embezzlement. In 2007 he spent two months in a remote prison in Manaus awaiting sentencing before he was released on appeal. Van Roosmalen claims assassins have been sent to the rainforest to kill him. Despite his legal struggles, he has described several new species that are now accepted. Others remain disputed.

In 2012 the ICZN Code was amended to include the principle of priority. Under its rule, whatever name was first assigned to a new species is the name that stands. Van Roosmalen believes that gives him the right to name the tapir. Meanwhile mammalogists like Helgen and Voss don't believe the little black tapir even exists. In 2014, Van Roosmalen filed an official complaint with the ICZN to confirm the name *T. pygmaeus* as valid under the principle of priority. Case No. 3650: the ICZN expects to rule on it soon.[12]

In the meantime, Cozzuol's work continues — like a tropical Captain Ahab, he squints into the shadows between the jungle walls. Soon, he says, he will embark on the first of several expeditions to field locations in the vast Brazilian interior. In a few weeks he'll travel to the Roosevelt River in Rondônia State. Shortly after Roosevelt charted the River of Doubt, it was renamed in his honor. It's the only river in Brazil named after a North American, which is a measure of the respect Brazil had for Roosevelt at the time. Cozzuol will survey the river's hairpin turns for telltale signs of the tapir in the wild. Later still, he plans to go to Pará State, an Amazonian region in northern Brazil, and search scientific collections there for more tapir skulls.

Back in the rainforest, Cozzuol swats another mosquito and looks into the tangled forest canopy above, like a green net closing. It's the same intertwined canopy Roosevelt peered into in 1914. It's the same place Van Roosmalen stalks, too. Early in the morning the jungle is quiet. The clearing is crisscrossed with well-used trails, worn down in places to mud packed like concrete. Tapirs, says Cozzuol, follow the same routes every day, like cattle. If he can find their tracks on the rainforest floor, he can find the tapirs.

He swigs from a water bottle, parts the vines, and steps into the darkness, listening for the sound of little black tapirs moving unseen between the trees.

4

A Taxonomic Confusion:
The Saki Monkeys
(*Pithecia* genus)

Several years ago, in 2001, Laura Marsh craned her neck and peered up into the green, misty undercanopy of the Ecuadorian rainforest. High in the air, indistinct dark shapes were moving between the branches—fast-moving shadows with long, bushy tails: saki monkeys. For several months Marsh had been putting together a biodiversity survey at the Tiputini Biodiversity Station—a 1,500-acre site in the eastern corner of Ecuador, in pristine lowland Amazon rainforest. It's a biodiversity hot spot. "I went and did surveys for two or three years," says Marsh. "I kept running across the saki monkeys, which I had never seen in the wild before."

Above her in the midstory, the shapes moved around in the mist, almost completely obscured by foliage. Agile and covered with long hair, the saki monkeys are neotropical primates belonging to the *Pithecia* genus. When fully grown they are about the weight of a half-grown domestic cat—about four pounds. They live their entire lives in the branches. "They're these little running, leaping, chirping things in the tops of the trees," says Marsh. At the time, there were five recognized species of saki monkeys, three of them polytypic— they each had subspecies, sometimes several. As the months passed at Tiputini, Marsh began to study the monkeys in earnest. "I did what any good birdwatcher would do, which is go and look them up, and read every thing about sakis," she says. But the monkeys weren't in the literature. They were undescribed. "Ah, shit," Marsh remembers thinking. "If it's a new species, I'm not a taxonomist: I'm not equipped to deal with this."

Thus began a decade-long project to revise the entire *Pithecia* genus. In 2014 Marsh published "A Taxonomic Revision of the Saki Monkeys *Pithecia* Desmarest 1804," in the journal *Neotropical Primates*.[1] For her research, Marsh relied heavily on archived monkey specimens in natural history collections worldwide, visiting thirty-six museums in twenty-eight cities across seventeen countries. In total she examined the preserved skins of 876 individual saki monkeys and 690 cleaned skulls. "I pretty much saw every dead saki there is to see," she says.

Old saki monkey specimens in museum collections often don't resemble their living versions at all. Many of the specimens Marsh examined were almost two hundred years old, grown brittle with the passage of time. Some look more like roadkill—flattened and misshapen. Others are strangely deflated, like shrunken balloons, with cotton balls visible in empty eye sockets. In some cases archived specimens have been prepared and mounted, set into unnatural poses, their unblinking glass eyes staring into space. One holds a fake pear. Others are frozen midleap. But to researchers even battered old specimens are a mine of information.

"I was one of the scientists thinking, 'You couldn't pay me enough to sit in a dusty museum and look at that stuff,'" says Marsh, but she quickly began to understand the untapped power of collections. With them, she can answer specific questions that aren't answered by studying animals in the wild. "I ended up going around the world," she says. "There are lots of collections—there's Boston, there's LA—that might have maybe two, five, or ten skins, but I was going for the ones that were in the double to triple digits for skins. And collections that had types."

The American Museum of Natural History in New York has 183 saki skin specimens. The Museu de Zoologia in São Paulo, Brazil,

111 skulls. The collection at the Muséum National d'Histoire Naturelle in Paris includes specimens collected as early as 1808. In Rio de Janeiro, Marsh measured saki specimens collected by Theodore Roosevelt. To collect more data, she studied illustrations painted by the early naturalists during expeditions. When possible, she observed live monkeys in the wild too. She visited zoos. She even used grainy photos of saki monkeys taken by tourists in the Amazon rainforest.

In 1785 the French naturalist Georges Louis Leclerc Buffon included an early description of the saki monkey in *Natural History, General and Particular*: "The Saki, which is commonly called the fox-tailed monkey, because its tail is garnished with very long hair, is the largest of the sagoins."[2] At the time, sagoin was a term broadly used to describe South American monkeys without prehensile tails. Buffon was referring to the white-faced saki, a species found across French Guiana, Suriname, and parts of northern Brazil. The *Pithecia* genus didn't exist. Two decades earlier, Linnaeus had given the white-faced saki its first Latin scientific name, the binomial *Simia pithecia*. "When full grown," Buffon wrote, "it is about seventeen inches long; but the largest of the other five species exceeds not nine or ten."

In other words, Buffon understood early on that there were several species. The *Pithecia* genus was erected by Desmarest in 1804 to house them, says Marsh. The earliest specimens she examined were collected about this time.

"The very earliest specimen I have is out of the Paris collections," she says. It was collected in 1808. "It's the type specimen for *Pithecia monachus*. It's this rubbishy-looking little juvenile that looks like a squirrel, basically. It's ridiculous. Frankly, I kept it for the sake of history. If I were a proper taxonomist—and had more balls, really—I would have just chucked that out and said, 'Pretend it doesn't exist,' because it was just so terrible." It was collected by French naturalist Étienne Geoffroy Saint-Hilaire. Adding to the confusion, in its original 1813 description, its type locality was recorded simply as "Habite . . . Le Brésil?" "For the next two hundred years—because it's such a crappy type specimen—nobody could tell what the hell it was," says Marsh.

There are now sixteen species of saki monkey. In all, Marsh named five new species, elevated three subspecies to full species status, and reinstated three other variants that had long been abandoned by taxonomists. It took years, and hundreds of specimens—Brazil, Sweden, Berlin, and elsewhere—to untangle centuries of taxonomic error and confusion. The errors had begun to multiply the moment the first European naturalist parted the foliage, stepped into the rainforest, and collected a saki monkey.

Many of the new species differ from one another in important ways, sometimes conspicuously: the black face of *P. rylandsi* peers out from a gravity-defying white bouffant. *P. pithecia* is almost a negative image of *P. rylandsi*—almost completely black with two white facial disks, a distinct half moon on each cheek. The holotype for *P. inusta* is an illustration made by Johann Baptist Spix in 1823 of a specimen he collected in 1817, during an expedition to Brazil. The specimen itself is now lost. Marsh described *P. cazuzai* from just three specimens—one of them collected in 1927. In most cases each of the new species occupies a distinct circumscribed habitat, overlapping one another in some regions but mostly separated by large watersheds that divide the Amazon basin into different parts, like a jigsaw puzzle. For millions of years the species have existed in their enclosed habitats, slowly diverging from one another. A few species are found across a wide swath of land that crosses borders. Others are confined to a tiny wedge of remote jungle, hemmed in on all sides by the rivers that cut through the trees. So far, *P. pissi-nattii* has been found only within in a fifty-mile-wide region of Middle Amazonas bordered on all sides by waterways. As early as 1852, English naturalist Alfred Russel Wallace had noticed the same forces at work. In an article titled "On the Monkeys of the Amazon," he proposed a still-held theory, that riverine barriers isolated the different monkey populations enough to give rise to new species: "The Sloth Monkeys, forming the genus *Pithecia*, have an extensive range as regards the genus, but the separate species seem each confined in a limited space."[3]

Those distributions were defined by the rivers. "During my residence in the Amazon district," he wrote, "I took every opportunity of determining the limits of species, and soon found that the Amazon, the Rio Negro and the Madeira formed the limits beyond which certain species never passed. The native hunters are perfectly acquainted with this fact, and always cross the river when they want to procure particular animals, which are found even on the river's bank on one side, but never by any chance on the other."

Now widely accepted, at the time it was a new idea—revolutionary.

When Marsh began the project, the *Pithecia* genus was in disarray. Primatologists had known as much for a while.[4] "They all knew sakis were a disaster," she says. In part this is because of their morphology—and because of human error. "The saki monkeys are extraordinarily variable," says Marsh. Often juveniles vary so much in appearance that they look totally different from adults of the same species for most of the time. A specimen collector might have shot a juvenile saki monkey from the trees in Guiana, or Ecuador, or northern Brazil and assumed, because it looked so distinct, that it represented a unique species. Those specimens slowly made their way from South America and were placed in collections around the world—seeds of error that bloomed and sent out tendrils everywhere. For centuries, primatologists have continued to refer to them. Together, misidentified and erroneously named specimens form the foundation of saki monkey taxonomy. But they are meaningless. The locality data were often wrong too, as with the *P. monachus* specimen St. Hilaire had collected in 1813—a sickly looking juvenile labeled "Habite . . . Le Brésil." After an expedition to the Amazon, Wallace had already begun to understand the effect inaccuracies would have on any attempt to map the distributions of different sakis. In "On the Monkeys of the Amazon" he wrote: "In the various works on natural history and in our museums, we have generally but the vaguest statements of locality. . . . S. America, Brazil, Guiana, Peru, are among the most common; and if we have '*River Amazon*' or '*Quito*' attached to a specimen, we may think ourselves fortunate to get anything so definite: though both are on the boundary of two distinct zoological districts, and we have nothing to tell us whether the one came from the north or south of the Amazon, or the other from the east or the west of the Andes."

Collectors field-dressed the misidentified specimens, salted the pelts, rolled them into sticky balls, and stuffed them into their backpacks. Eventually they sent the pelts back to the American Museum of Natural History in New York City and elsewhere. "There are a lot of *Pithecia* skins in the American Museum that look like foxes," says Marsh.

Iquitos is a sprawling port city of almost 400,000 inhabitants in northern Peru. Its buildings and wide roads cling to a slow bend in the Amazon River. To the east, the streets end where the water begins; to the west and south, the roads taper off, terminating at the edge of the rainforest, which looks like a green wave about to engulf the city. Marsh traveled to Iquitos to examine the saki specimens at the Museo de Zoología collection, which is part of the Universidad Nacional de la Amazonía Peruan. The museum collection has been battered by the elements.

"It's open air," says Marsh. "They have sacks and sacks and sacks and sacks of skulls. They can't really keep skins because it's right in the middle of the Amazon, and the humidity eats everything."

Marsh had to open each sack, digging through an assortment of skulls to find the saki skulls inside. "It's a collection waiting to be curated," she says. The problems Marsh saw in the Iquitos collection are common in the developing world, but not confined to it. The collection at the Zoologische Staatssammlung in Munich is endangered too.

"Their main collection—hundreds of years of fantastic data—is pelts in stacks, just on the ground," she says. "They don't have the money or the staff to go through it." Curator Richard Kraft pointed at a pile of pelts on the ground and told Marsh they were the type specimens. Then he turned and pointed toward a different pile of skins—a pyramid of old fur—on the ground, saying, "If you want the rest of the collection, you'll have to dig in that pile." Centuries of human endeavor had been reduced to a mound of preserved skins piled on top of each other on the floor. "I'm five-foot-six," says Marsh. "It was six or eight inches taller than me."

During her research, Marsh tried to see living representatives of every species of saki monkey included in the revision. There was one species that evaded her: *Pithecia vanzolinii*—Vanzolini's bald-faced saki. No other saki monkey even closely resembles it. It was previously a subspecies, but Marsh elevates it to species status. The holotype was collected in 1936. It's part of the collection at the Museu de Zoologia in São Paulo, along with several paratypes collected around the same time. "I've seen every specimen of them," says Marsh. "They're very distinct—very different. But nobody's seen them alive since the 1930s."

The specimens all came from one place: a remote region of the Brazilian rainforest near the Peruvian border. A dangerous place, says Marsh, who is planning an expedition to find the monkey in the wild. In 2014 the Brazilian government's Indian Affairs department, FUNAI, reported that an estimated seventy-seven uncontacted tribes still exist in the Amazon rainforest, cut off from the modern world. The Alto Rio Juruá region, in Acre State, where the paratypes of *P. vanzolinii* were collected, is home to some of them. They are pressed in on all sides by loggers and mining operations, by the advance of modernity: by destructive agricultural practices and the sudden appearance of new settlements, by the arrival of electricity and modern weapons, and by the illicit movements of violent drug cartels.

"There hasn't been a mammal survey into that watershed in about one hundred years," Marsh says. "Some of the most recent expeditions have been for plants and reptiles and birds and stuff—but that's only been in the last fifty years. It's pretty well unknown."

As recently as June 2014, new footage surfaced of uncontacted indigenous tribes, filmed from the air by a BBC crew: several people standing in a well-trodden red earth clearing in the forest. They stand by a long, low thatched hut. A spindly column of blue smoke rises into the air between well-spaced cultivated banana plants. Their bodies painted blood red with annatto, the Indians raise their bows, aiming at the airplane as it banks above the trees. If anyone has seen *P. vanzolinii* since it was last collected in 1936, it is these people: the uncontacted ones.

"I'm just crossing my fingers that we don't run across any of those guys," says Marsh. She plans to live on a houseboat with her expedition team, dispatching skiffs and drones into the jungle, searching the tops of the trees for the monkeys. "As beautiful and romantic and dreamy as it all sounds, I really don't want to be shot full of a bunch of arrows and have my boat taken from me," she says. "They're uncontacted for a reason."

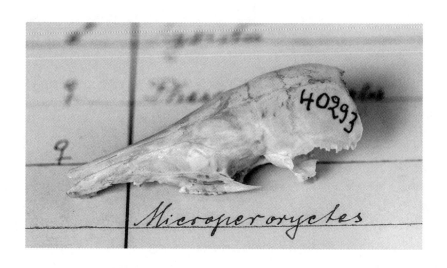

5

Scattered to the Corners of the World: The Arfak Pygmy Bandicoot (*Microperoryctes aplini*)

"It's the world's smallest bandicoot," says Kristofer Helgen from his post at the University of Adelaide Environment Centre. He's talking about a novel species of pygmy bandicoot, a small, light-footed marsupial from New Guinea with a long, tapered snout, which he named in 2004. "You can't really compare bandicoots to any other kind of mammal that's not a bandicoot," he says. "They're only in Australia and New Guinea; some people say they're a little bit like an Australian version of a rabbit."

In fact it looks like a Dr. Seuss mash-up of a mouse, a shrew, and a rabbit, with a long thin tail and a bold black dorsal stripe along the middle of its back. But the Arfak pygmy bandicoot hasn't been seen alive for decades—at least no field biologists have reported a sighting. Four known specimens exist. They all came from the same tiny, well-defined region: a mountainous part of the Bird's Head Peninsula, in the Indonesian half of New Guinea, near the shoreline of two remote mountain lakes ringed by steep green hills. For a long time the Arfak pygmy bandicoot was mistaken for another species of small bandicoot also found on New Guinea. But, says Helgen, it's quite distinct from it—and from all other species of bandicoot.

"They're dainty," says Helgen. "They're beautiful. This is a tiny and beautiful, charismatic animal."

New Guinea is a large Melanesian island—slightly larger than the state of Texas—to the north of Australia. It contains the world's third largest area of rainforest after the Amazon and Congo. Within the forest there are numerous tribes of indigenous people—more than a thousand languages are spoken there. Some people say the island is shaped like a bird of paradise. Perched in the ocean, the bird looks west across Indonesia, toward Malaysia. The Bird's Head Peninsula is at the western end of the island in New Guinea—the half of the island that is Indonesian—and the Bird's Tail Peninsula is to the east, in the Independent State of Papua New Guinea.

New Guinea is recognized as a biodiversity hot spot. Many of its species are endemic—they are found only there. Between 1998 and 2008, according to a World Wildlife Fund report published in 2011,[1] more than a thousand new species were discovered in New Guinea. In 2009 an international team of zoologists, including Helgen, surveyed Mount Bosavi, a dormant volcano in Papua New Guinea. Two and a half miles wide and more than half a mile deep, the almost inaccessible crater was home to an array of species, including more than forty that were entirely new to science: fanged frogs, nectar-eating bats, camouflaged spiders, iridescent beetles, and a large woolly rat weighing more than three pounds that is still unnamed.[2]

Situated in the northeastern part of the Bird's Head Peninsula is the Arfak mountain range, crowned by the green peak of Mount Arfak, at an elevation of 2,955 meters. In 1858, when English naturalist Alfred Russel Wallace visited New Guinea, he described the Arfak mountains thus: "There is a continued succession of jagged and angular ranges of hills, and everywhere behind them, ridge beyond ridge stretch far away into the interior. Over the whole country spreads an unvarying forest, of a somewhat stunted appearance, broken only by the very widely-scattered clearings of the natives on the lower slopes."[3]

On Google Earth, the Bird's Head Peninsula is a wide green place. A small fleet of white clouds drifts above it like empty speech bubbles. To the southwest, a pan-flat coastal plain is punctuated with a procession of river deltas that curlicue southward, emptying into the ocean along the length of the coastline—blue rivers narrowing, snaking, and dwindling to the north, as delicate as calligraphy. It is wet and riverine, divided and bifurcated by the irregular, serpentine tracings of a hundred waterways.

But to the northeast the rumpled mountains rise suddenly, tight against the coastline. It is montane forest—dark green in the shadowy pleats of the mountains. In a flattened swale sit twin mountain lakes. *Microperoryctes aplini*, not much larger than a mouse, lives in the montane forest near the lakes.

In 1928 the famous biologist Ernst Mayr was in New Guinea, making his way through the Arfak Mountains with his guides, trailed by a small army of overloaded porters.[4] Later Mayr would go on to redefine evolution, writing numerous influential books on the subject, but at that time he was twenty-three. It was his first expedition.

In a photo taken during the expedition, Mayr looks like a schoolboy, half sitting on a skinny fallen tree in front of a thick bank of ferns. He's wearing expedition gear and gaiters, with a floppy-brimmed hat.

Primarily, Mayr was in New Guinea to collect bird specimens. He had been dispatched to the island by Lionel Walter Rothschild—Lord Rothschild, the second Baron de Rothschild. Portly, with an elaborately waxed handlebar mustache and a thick beard, Rothschild was an eccentric member of the English aristocracy and heir to the Rothschild and Sons banking fortune. In 1889, at age twenty-one, he had gone to London to work for the family-owned international banking firm, but without much passion. He was obsessed with natural history instead. Famously, Rothschild drove to Buckingham Palace in a carriage pulled by zebras to prove they could be trained like horses. In his lifetime he amassed an enormous scientific collection—the largest private collection in history: more than 300,000 bird skins, 200,000 bird eggs, 2,250,000 pinned butterflies, 30,000 beetles, and many mammals, reptiles, amphibians, and fish.[5]

By 1892 he housed the collection in a museum of its own, which was open to the public and became the Natural History Museum at Tring, in Hertfordshire, England. Eventually Rothschild retired from the banking industry and dedicated the rest of his life to maintaining his natural history collection. To expand his collection, he employed some of the leading biologists of the era, including entomologist Karl Jordan and ornithologist Ernst Hartert. By the early 1930s, crippled with debt, Rothschild sold most of his collection to the American Museum of Natural History. When he died in 1937, what was left of it became part of what was then the British Museum of Natural History collection.

About the time Mayr was dispatched to New Guinea—and for several decades before that—fashionable women in American and European cities wore hats decorated with elaborate, bright-colored plumes. The feathers came from highly ornamented species like snowy egrets, which had been hunted almost to extinction by the end of the nineteenth century. Feathers from several bird-of-paradise species were used too. Inevitably, intrepid plume hunters turned to New Guinea, a relatively unexplored island with the greatest diversity of birds-of-paradise in the world. To satisfy the global fashion market, bird-of-paradise feathers were harvested by the millions from the wild and sent to New York, London, and Paris. Occasionally, conspicuous among the feathers were unknown varieties—strange new forms belonging to unidentified species. An avid ornithologist, Rothschild was spellbound by the unfamiliar feathers. He arranged to buy unidentified plumes directly from dealers, then sorted through them in safety—in Europe.

But in 1928 he gave Mayr a different directive: go to the source. Go to New Guinea and collect unknown birds-of-paradise from the slopes of the Arfak Mountains; scour the mossy forests of the Wandamen and Cyclops Mountains for the undescribed species. Mayr wasn't the first scientist to go in search of them. French naturalist René Lesson had gone to New Guinea in 1824 and documented birds-of-paradise in the wild for the first time. When he saw one in the trees, he reported that he was too astonished to shoot it: "It was like a meteor whose body, cutting through the air, leaves a long trail of light."

In fact, during the Victorian era New Guinea had seen the arrival of one ornithological expedition after another. Everyone was obsessed with the search for unknown birds-of-paradise. When biologist Alfred Russel Wallace visited in 1858 he was searching for them too. He'd written extensively about birds-of-paradise. On Halmahera in the Maluku Islands he described a medium-sized bird with an emerald green breast shield and two pairs of long white plumes falling from its shoulders. Later George Robert Gray of the British Museum named it Wallace's standardwing (*Semioptera wallaci*) in his honor. But in New Guinea he was unsuccessful. In total, Wallace made five voyages to different parts of the island and saw only five of the thirteen bird-of-paradise species known to live there. He discovered no new species on the island.

"Nature seems to have taken every precaution that these, her choicest treasures, may not lose value by being too easily obtained," he wrote. "First, we find an open, harbourless, inhospitable coast exposed to the full swell of the Pacific Ocean; next, a rugged and mountainous country, covered with dense forests, offering its swamps and precipices and serrated ridges an almost impassable barrier to the central regions; and lastly, a race of the most savage and ruthless character, in the very lowest stage of civilization."

Next was Hermann von Rosenberg, a German naturalist who arrived in 1869, collecting a huge number of birds. He was followed by another German—Adolf Meyer, from Dresden. Then in September 1872 the Italian explorer and naturalist Luigi D'Albertis arrived. D'Albertis had an extravagant beard that stretched in a bushy fork halfway down his chest, like a strange black pitchfork. On a later expedition he became the first person to chart the Fly River—seven hundred miles to the southeast, on the border between New Guinea and Papua New Guinea. Aboard an Australian launch called the *Neva*, he made his way almost six hundred miles upstream, constantly firing rockets from the side of the boat to keep the hostile and aggressive natives—known cannibals—from attacking. Standing before a crowd of Yule Island natives in 1874, D'Albertis held

aloft a shell filled with burning methylated spirits and threatened
to set fire to the ocean.

The Italian kept a detailed log:

September 1872: On the 20th I had still some rice, but only sufficient for
two days. My salt was finished. Since the 18th I had reduced the daily
ration of rice to half the usual quantity, but I managed very well to subsist
on yams, and the birds and other animals that I shot, so that I had often
a novel and very excellent meal of roasted Birds of Paradise, of the rarest
kinds, the beautiful and brilliant skins of which I had previously prepared
for my collection. The water was so good that it did not require brandy to
qualify it for drinking, but the want of salt I felt very much.

In fact, during his difficult months in the pathless tropical forests of New Guinea, he obtained many specimens.

D'Albertis again:

Among my collection of insects there are a number new, and I found some
very fine specimens of Cetonia, and Melolontha. Mammalia are compara-
tively rare, for in all the parts of New Guinea at present known there are
only three species of Cuscus, a Perameles, Papuan Wild Hog (*Sus papuensis*), two or three species of Dendrolagus or Tree Kangaroo, a fruit-eating
Bat, an animal resembling a striped Phalanger (*Dactylopsila trivergata*), a
species of squirrel, two or three species of mice, and no bats from New
Guinea having yet been described, there is [*sic*] probably six or seven
new species in my collection.[6]

Somehow the Arfak pygmy bandicoot had managed to evade
everyone.

In 1928, when Mayr departed for New Guinea, an expedition to
a mostly unknown place was still an incredibly dangerous under-
taking. Rothschild had already lost plenty of collectors: William
Doherty had died of dysentery in Nairobi in 1901; George Ockenden was lost to typhoid fever in 1906, in the Peruvian Andes.[7] In
1921 the bird collector Noël van Someren was gored to death by

a buffalo. Three collectors had died of yellow fever in the Galápagos Islands. In Silchar, India, E. C. Stuart Baker survived a leopard attack, but his left arm had been bitten off. Then there were the locals: in 1901, when English lepidopterist Antwerp Pratt (Antwerp had a sister called Vienna) arrived to collect butterflies in the Arfak Mountains, the natives were so aggressive that he fled east to Papua New Guinea and collected specimens there instead.[8]

Somehow Mayr managed to avoid conflict with the natives—who were always at war with each other—and he remained healthy. "In the interior," he said in a 1997 interview, recorded when he was ninety-three, "I came to a number of villages where no white person had ever been, and it was really an untouched nature. The flowers were also most overwhelmingly rich and astonishing. The bird life was incredible."[9]

In all, between April and October 1928, he collected about 2,700 bird skins in New Guinea—representing more than 350 species and subspecies. But he was collecting specimens from other orders too. In the Arfak Mountains and elsewhere, Mayr collected 260 mammal specimens. Among them was a small brown bandicoot with a long dark band running along its back. Helgen named the new species *Microperoryctes aplini* in honor of Ken Aplin, an Australian zoologist who has spent decades trying to resolve bandicoot taxonomy.[10]

All together, there are only four known specimens. The oldest—the bandicoot Mayr collected in 1928—now resides in Berlin, at the Museum für Naturkunde. Its skull is fragile and broken in places. As soon as Mayr returned to Europe, a catalog of the numerous bird species he collected in New Guinea was published, but no one really cared too much about the mammals he'd brought back. The specimen Helgen designated as the holotype was collected in 1963. Its delicate and elongated, almost bird-like, skull is part of the mammal collection of the Bernice P. Bishop Museum in Honolulu, along with its preserved skin. In addition, Helgen found two more specimens: "One was in the Bogor Museum in Java—in Cibinong, in Indonesia, and the last one was in the National Museum in Papua New Guinea."

Therein lay the reason for the delay in its description. Taxonomists had suspected for some time that the specimens represented
a new species. A photograph of one had even been included in *Mammals of New Guinea*, a 1995 field guide by prominent Australian
mammalogist Tim Flannery. Incorrectly labeled *Microperoryctes
murina*, the photo showed a walnut-colored pelt with a darker band
down its center. The specimen was clearly different from known
bandicoot species. In order to describe it, somebody had to actually examine them all.

   "These things were literally scattered to the corners of the
world," says Helgen. No two specimens are even on the same continent. "Anyone who had ever puzzled over whether this was a new
bandicoot probably only ever had a chance to view one of them.
If you travel, if you can see all these—and if you can see them all
in short succession—you realize these four things have a totally
unique set of skull characteristics. It has a unique color pattern ex-
ternally. You see it for what it is: it's a bandicoot that has never got
a scientific name."

   Including *M. aplini*, the *Microperoryctes* genus of bandicoots consists of five named species belonging to the Peramelemorphia order
of marsupial omnivores. In New Guinea they share their habitat
with a couple of species of tree kangaroos and echidnas, an array
of moss mice, and several other small mouselike marsupials—the
narrow-striped marsupial shrew (*Phascolosorex dorsalis*), the three-
striped dasyure (*Myoictis melas*), and others. Since its collection in
1928, the Mayr bandicoot specimen had been mistaken for another
species of bandicoot, *M. murina*—the mouse bandicoot. Small like
the Arfak pygmy bandicoot, the mouse bandicoot is distinctly different and has what Helgen calls an "inornate, smoky-grey pelage,
which is soft and woolly like that of the shrews and moles."

   M. murina lives farther to the west in the moss forests of the
Weyland range—another hilly formation distinct from the Arfak
Mountains. It has been found only there, in a tiny, limited range,
and is also known by only three specimens. More recently, Helgen
has turned his attention to another series of bandicoot specimens
in the American Museum of Natural History collection. They came

from the Snow Mountains, an extended saw blade of peaks in the Central Cordillera of New Guinea. Collected during a 1938 expedition, the bandicoots likely represent another unnamed species that has gone unnoticed until now.

Helgen suspects the Arfak pygmy bandicoot still thrives in the mountains of northwest New Guinea. According to a 2012 phylogenetic study,[11] *M. aplini* has likely lived in the montane forests of New Guinea, distinct from its closest relatives, for at least several million years. It diverged from its closest named relatives an estimated four million years ago. And, says Helgen, it hasn't gone anywhere since it was last collected—one of the museum specimens is from 1986. Most of the specimens were obtained near the shoreline of the Anggi lakes, in a narrow elevational band between 1,890 and 2,200 meters.

"All someone has to do is make a trek up to those lakes, and spend a few days trapping bandicoots," says Helgen. "It's a pretty small animal, so it's going to be there. I just haven't gone to find it yet."

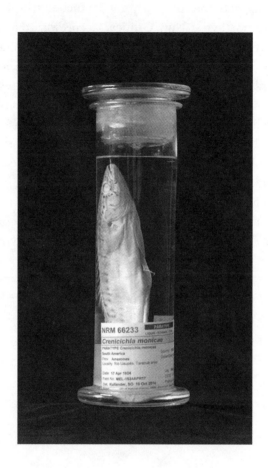

NRM 66233

Crenicichla monicae

PARATYPE Crenicichla monicae
South America
Prov. Amazonas
Locality Rio Uaupés, Taraçua area

Date 17 Apr 1934
Field No. MEL-1934APR17
Det. Kullander, SO 19 Oct 2014

6

The One That Got Away for 160 Years: Wallace's Pike Cichlid (*Crenicichla monicae*)

On the morning of Friday August 6, 1852, Alfred Russel Wallace was summoned to the deck of the brig *Helen*. The boat was in the middle of the Atlantic Ocean, and Wallace had already been at sea for twenty-six days. He was used to hardship. He'd spent the previous four years in the Amazon rainforest, exploring uncharted territory and collecting natural history specimens for his own collection and for museums back home in England. The hold was filled with his valuable specimens—many were new to science and irreplaceable. The twenty-nine-year-old Welsh-born naturalist had even stowed several live specimens: there were parrots and parakeets, some monkeys, and a wild forest dog on board.

The captain said to Wallace, "I'm afraid the ship's on fire. Come and see what you think."

Ten months earlier, in the depths of the rainforest, Wallace had contracted a fever that almost killed him. Still sick, he now stood with Captain Turner on the deck of the *Helen* and watched as smoke billowed from the forecastle. The small crew frantically threw buckets of water into the hold, but the fire was unstoppable and engulfed the ship. The captain gathered up his chronometer, sextant, compass, and charts, and the crew began to prepare the rescue boats: a longboat and the captain's gig.

"I got up a small tin box containing a few shirts," Wallace recounted in a letter to his friend, the botanist Richard Spruce, "and put in it my drawings of fishes and palms, which were luckily at hand."

The boats were quickly loaded with supplies: "Two casks of biscuit and a cask of water were got in, a lot of raw pork and some ham, a few tins of preserved meats and vegetables, and some wine." Then the men clambered aboard. Gripping a rope to lower himself, Wallace slipped, stripping the skin from his hands, and fell into the boat. In time the *Helen* sank to the bottom of the Atlantic, taking with it an unknown diversity of new species Wallace had collected.[1]

Finally, in September 2015 Sven Kullander, an ichthyologist at the Swedish Museum of Natural History in Stockholm, named one of the undescribed species—a distinctively patterned red fish. Wallace had collected the specimen from the Amazon basin in 1852. Kullander has given it the scientific name *Crenicichla monicae*.

"*Crenicichla monicae* belongs to a group known as pike cichlids," says Kullander. All together, the genus includes almost a hundred known species distributed across tropical and southern South America, east of the Andes. "Pike cichlids are elongate," he says, "and most of them have a pointed snout and large mouth, reflecting their carnivorous habits."

The fish has a long, spiny dorsal fin that runs almost the length of its body, and it grows to a length of about ten inches. Above its lateral line, its long, narrow body is speckled and patterned with distinctive dark spots. Kullander's descriptive paper, published in *Copeia*, is titled "Wallace's Pike Cichlid Gets a Name After 160 Years: A New Species of Cichlid Fish (Teleostei: Cichlidae) from the Upper Rio Negro in Brazil."[2]

A major tributary of the Amazon, the Rio Negro is a blackwater river that begins in southern Colombia and hairpins southeast through the rainforest. At Manaus, the Negro and the Amazon meet. Wallace had collected the fish upstream, on the swift upper reaches of the Rio Negro, sometime before 1852. At least one specimen of *C. monicae* was in the hold of the *Helen* when she sank, says Kullander. He knows this because of a pencil drawing Wallace made after he collected the fish. He'd worked as a surveyor and was trained to draw with great accuracy. As the ship burned and began to list, Wallace grabbed his drawings of fishes and palms and ran to escape the flames. When he fell into the boat, the drawings fell in with him.

A world expert on cichlids, Kullander is in his middle sixties and has a long silver ponytail and pale blue eyes. He's been studying fish for more than fifty years. When he was still a high school student, he was given his own research space.

"I tried birdwatching, but it was boring," he says. Instead, he settled on fish. In the 1970s Sweden saw a boom in the importation of African lake cichlid species as aquarium specimens. Suddenly more and more species were available, and some of them were uncommon forms. As a teenager Kullander took advantage of the opportunity and became an avid aquarist, maintaining tanks filled with exotic cichlids. "I tended more to the neotropical forms; however, I was very familiar with all the literature and kept all the species I could find," he says. "I maintained a large correspondence with other hobbyists and scientists and neglected school as much as possible."

He wrote his first academic papers without supervision as an undergraduate. Finally, during his graduate studies, Kullander was supervised by Bo Fernholm—at the time, the only fish taxonomist in Sweden. In all, Kullander has named more than a hundred new cichlid species. Worldwide, he says, there are almost two thousand known species of cichlids, found in Asia, Africa, and North, Central, and South America. Among other diagnostic criteria, ichthyologists identify new cichlid species by subtle differences in the pharyngeal teeth—two plates in the throat covered in beds of toothlike spines that come together to grind up food.

Charles Darwin published *On the Origin of Species* in 1859, but he didn't form his theory of evolution in isolation. He had collaborated with people like Wallace, a quiet, bespectacled man who was the first to develop some of the fundamental theories of biogeography. It was Wallace who seemed to grasp intuitively the importance of geographic isolation of organisms in the process of speciation, and he shared those ideas with Darwin as they corresponded. For years Wallace had explored the darkest parts of the Amazon rainforest, collecting specimens like *C. monicae*. Later he also spent time in the Malay archipelago, where he demarcated the Wallace Line—a theoretical boundary that runs northward through Indonesia and between Borneo and Sulawesi before veering northwest past the Philippines. Vastly different species are found each side of the line: Australasian species to the east, Asiatic species to the west. In places the barrier between the two ecozones is nothing more than a narrow channel of water. With very few exceptions, species are found either on one side of the line or on the other. Occasionally, in a region known as Wallacea, the line is less clear.

Back in 1852, Wallace and the other men sat in the rescue boats in the middle of the Atlantic Ocean and watched as the flames rushed across the deck and up the sails, consuming what was left of the *Helen*. "Soon afterwards," he wrote, "by the rolling of the ship, the masts broke off and fell overboard, the decks soon burnt away, the ironwork at the sides became red-hot, and last of all the bowsprit, being burnt at the base, fell away also."

Night came. The men in the boats stayed close to the burning ship as flames illuminated the black water. Sometime in the night the *Helen* rolled over in the waves—a great hissing cauldron of fire, the cargo burning in a liquid mass at the bottom. Wallace was calm; he expected to die. As the fire sputtered into the ocean, he sat in the boat thinking of his lost specimens: the forest dog; the jars filled with fish; the insects and parakeets; three woolly monkeys. "I had taken some trouble to procure and pack an entire leaf of the magnificent Jupaté palm (*Oredoxia regia*), fifty feet in length, which I had hoped would form a fine object in the botanical room at the British Museum."

He stared at his hands, still raw and stinging from rope burns, and enumerated his specimens. "All of my private collection of insects and birds since I left Para was with me," he wrote, "and comprised hundreds of new and beautiful species, which would have rendered (I had fondly hoped) my cabinet, as far as regards American species, one of the finest in Europe."

It was all gone, but Wallace's illustrations survived. Eventually they became part of the collection at the Natural History Museum in London. In 2002 they were published as *Fishes of the Rio Negro*. Plate 194 is a finely detailed black-and-white pencil drawing of *Crenicichla monicae*, with its long, spiny dorsal fin. The mouth is slightly open, and the back and upper body are patterned with dark spots. The fish, Wallace had noted, was distinctively red. With a dull crimson band along its sides, even its eyes are orange-red.

In the intervening years, most of the other cichlids in Wallace's drawings had been identified — usually at least to genus and often to species. Many of the fish he drew were known before he collected them. They had been described in an 1840 monograph by Austrian ichthyologist Johann Jakob Heckel, who relied on specimens collected by Johann Natterer, a fellow Austrian who was part of an 1817 expedition to Brazil. He returned to Vienna almost twenty years later with an enormous collection of natural history specimens. Other species remained unknown. In 1989 Kullander described *Acaronia vultuosa*, another species Wallace had drawn in the Amazon. A few species waited even longer. Like the fish in plate 194.

In 1923 a group of Swedish biologists traveled to some of the same locations Wallace had visited seventy years earlier. There were three friends: Douglas Melin, Arthur Vilars, and Abraham Roman. Melin and Roman were biologists, and Vilars, an engineer, was their assistant.

From Manaus they ascended the Rio Negro northward. The river there is really a collection of channels traveling in the same direction. Where the Rio Uaupés meets the Rio Negro, they took the fork eastward along the Uaupés, into the darkness, tracing the tight turns of a swift brown river that grew narrower and more sinuous with every mile. They collected biological specimens along the way. By mid-1924 things had changed. Roman had left for Sweden in April. Vilars was dead by June — killed by a fever.

Melin then returned to Manaus, where he shipped specimens home before continuing as a one-man expedition in Peru. Together the biologists had collected thousands of specimens—frogs, catfish, jumping spiders, and numerous botanical samples. In April 1924 Melin and Vilars had collected several specimens of a distinctive spotted red fish from the Rio Negro drainage basin at Taracuá. It was the same fish Alfred Russel Wallace had collected and then watched sink into the ocean. Eventually, says Kullander, all the specimens from the expedition were sent to the Swedish Museum of Natural History.

"All together, the Melin fish collection was not a big one," he says. It comprises 130 jars, each containing one or more specimens stored in alcohol. "People going on expeditions today take back thousands of specimens," says Kullander.

Melin's specimens of *C. monicae* remained unidentified in Stockholm. Slowly the red scales faded to pink and eventually to pale yellow. The eyes turned milky and opaque. Molecules in the specimen broke down and began to degrade. In the 1950s, ichthyologist Otto Schindler visited the collection in Stockholm, saw the specimens, and took them to the Bavarian State Collection of Zoology in Munich, where he worked as a curator. Again they sat unidentified for decades. In the 1990s Kullander found the specimens—still in Munich.

"I found these spotted fish, which I recognized must be something new," he says. He took two of the specimens back to Stockholm but left a third one because it didn't have any color. With the constellations of dark spots that mark its upper body and its long, spiny dorsal fin, the fish that Melin and Wallace both collected is unique among cichlids. From the handful of specimens that exist, says Kullander, the markings seem to be present only on females.

"Most crenicichla species look very similar. . . . The distinctive color pattern of female *Crenicichla monicae* is very evident on Wallace's drawing and enabled us to identify it as the spotted species in Melin's collection." Compared with its closest relatives, its long dorsal fin has an additional spine. Its pharyngeal teeth are different in important ways. And "unlike the majority of pike cichlids," says Kullander, "in which the scales are rough to touch from small spines along the scale margin, *Crenicichla monicae* is one of three species in the genus that have smooth scales."

More than 160 years after Wallace collected it, then lost it— and almost a century after Melin collected it again at Taracuá— the fish has a name. Since Kullander described the fish, researchers have discovered other lost species Melin also collected during his time in the Amazon. In January 2016 Auburn University biologists Milton Tan and Jonathan Armbruster named a new species of suckermouth armored catfish from a single specimen Melin collected in the Rio Negro basin. *Hypancistrus phantasma* is a ghostly pale, broad-shouldered catfish with a small, underslung mouth and a wedge-shaped body. *Phantasma* means "phantom." Like Wallace's pike cichlid, it waited in a jar for almost a century. Melin and Vilars collected the holotype from the Rio Negro on February 14, 1924. But it hasn't been seen since.

In 1924 it existed flattened against the riverbed in the deep, swift water. Maybe it exists there still, ghostlike, a phantom fish. Perhaps there are other phantom fish too, suspended in jars in Stockholm and elsewhere. As the years pass the specimens continue to fade, slowly losing their colors and the distinctive patterns and striations that might once have set them apart from the others.

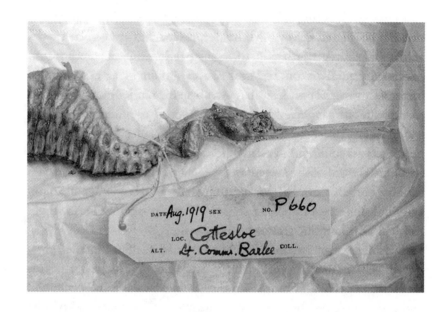

7

Here Be Dragons:
The Ruby Seadragon
(*Phyllopteryx dewysea*)

Faded after years in storage, the specimen is now the color of a ripening peach. Its surface is corrugated with ridges where the bony plates of its body meet one another. Its snout is elongated like a pipe. Described in 2015 and given the Latin name *Phyllopteryx dewysea*, the ruby seadragon is one of only three known species of seadragons — members of the Syngnathidae family — and the first novel species discovered since 1865.[1]

In life the body of the ruby seadragon is a deep, intense red — the color of a fresh, wet heart: *profondo rosso*. But this pale specimen is not alive. It washed ashore almost a century ago, in August 1919, and was found on the sand at Cottesloe Beach, a wide-mouthed bay on the coastline of Western Australia near Perth. A member of the Australian Royal Navy — Lieutenant Commander Barlee — found it there on the shoreline. Arterial red, it must have been conspicuous among the green tangles of seaweed on the slope of the beach. Barlee picked the seadragon up.

Eventually the specimen became part of the natural history collection at the Western Australian Museum in Perth, a few miles inland from where it was found. It has remained in the collection since then, a yellowing handwritten tag tied around its neck with string. It might have stayed unnoticed in the collection forever, but Nerida Wilson, a senior research scientist at Western Australian Museum (WAM), found it there, misidentified as a common seadragon.

October 10, 2007: a marine research vessel pitches on a heavy swell in the Recherche archipelago, a scattering of small islands off the south coast of Western Australia. Slightly east of Middle Island, a trawl net slowly emerges from the water, pulled aboard by a clattering winch. On deck, researchers from the University of Western Australia's Centre for Marine Futures are completing another biodiversity survey. Inside the net there are eagle rays, pink snappers, silver trevallies, and humphead wrasses. But the researchers have caught something else too. Trapped in the wet folds, trawled from a depth of about fifty meters, is a bright red seadragon with a deep S-shaped keel and a straight, pipelike snout like a clarinet.

"None of us were on that cruise," says Josefin Stiller, a graduate student and seadragon researcher in the marine biology program at Scripps Institution of Oceanography in San Diego. "If there had been someone on the boat who knew seadragons, that person would totally have spotted it—that it was weird—just because of the color." Instead, says Stiller, the researchers on board took a photograph of the seadragon and then froze its body, later accessioning it at the Western Australian Museum as a common seadragon. No one looked at the seadragon again until Stiller requested it. That first photograph, taken at sea on the deck of the boat, is included in the descriptive paper she published in 2015, in the journal *Royal Society Open Science*, with her coauthors Nerida Wilson at WAM and Greg Rouse of the Scripps Institution of Oceanography.[2]

There are two other known species of seadragon: the common seadragon (*Phyllopteryx taeniolatus*), first described in 1804, and the leafy seadragon (*Phycodurus eques*), described in 1865. Both are highly ornamented with leafy appendages that resemble kelp fronds, camouflaging the slow-moving fish to protect it from predators.

An early account of the leafy seadragon in *Fish and Fisheries of New South Wales* (1882) by naturalist Julian Edmund Tenison-Woods describes it like this: "It is the ghost of a seahorse, with its winding sheet all in ribbons about it; and even as a ghost it seems in the very last stage of emaciation, literally all skin and grief. . . . The odd thing about these strange fishes is that their tattered cerements are like in shape and colour to the sea-weed they frequent, so that they hide and feed with safety. . . . Just as the leaf-insect is imitative of a leaf, and the staff insect of a twig, so here is a fish like a bunch of sea-weed. If this is development, it stopped here only just in time; one step more and it would have been a bunch of kelp."[3]

Seadragons and seahorses are related, but there are distinct and significant genetic differences between them. Belonging to the genus *Hippocampus*, there are fifty-four known species of seahorse. They are more widespread and are found in shallow tropical and subtropical waters from North America to South America in the Pacific Ocean, and from Nova Scotia to Uruguay in the Atlantic. "Seahorses get discovered all the time," says Stiller. The same is not true for seadragons, which are found only along the south Australian coastline. "Discovering a third species definitely was something special," she says.

Stiller uses genetic techniques to research the diversity between and within different seadragon species across their overlapping ranges. "We want to understand how many populations there are in Australia in the wild, and how these populations are connected to each other — how genetically diverse they are," she says.

To do that Stiller needs fresh tissue from individual seadragons. For several years she and Rouse, collaborating with Wilson in Australia, have steadily been collecting samples to build a permanent tissue bank. "Most of them we caught them in the wild, taking little tissue clips," says Stiller. Those samples are then sequenced: their DNA is isolated and unraveled: decoded, interrogated. DNA sequences are then compared with one another and carefully assessed for differences. "I also always contact museums and ask if they have any tissue samples we could use," she says. "In most cases fishes will get thrown into formalin, so you can't really use them for DNA work."

When Stiller contacted the WAM, she was offered a sample from the red seadragon captured in October 2007 in Recherche. It was still in the freezer. Like the others, the tissue was processed, and its DNA was isolated and sequenced. "We have over 250 samples of seadragons, and it didn't match any of them," she says. "It was actually very, very distinct from the other two species genetically."

Stiller hadn't anticipated an outlier. It clearly didn't belong to either of the other known seadragon species. "I sat on the results for quite a while," she says. She checked her data, then rechecked it. Then she resequenced the DNA isolated from the tissue samples to confirm that it was different and that there was no contamination of the tissues. By then Stiller still hadn't actually seen the seadragon specimen from Recherche. It was a male, and it was pregnant. On the underside of its tail it carried a clutch of bright red eggs. Within each egg was a perfect miniature replicate of the seadragon, like a tiny red curled comma. Staff at WAM sent Stiller fifty of them.

"I took the eggs and sequenced all of the individuals," she says. They were all outliers too, like the seadragon they were harvested from. Growing more confident of her findings, Stiller requested collection details for the specimen and received the locality and the depth at which it was collected. Fifty-one meters: deep for a seadragon, thought Stiller. Members of the other two species of seadragon typically are found inshore and at much shallower depths. Most of Stiller's tissue samples were from seadragon specimens collected much closer to the surface.

"Then they said, 'Oh, we have this photo,'" says Stiller, "and they sent this photo to us—of this red fish." Suddenly everything became clear: the unexpected genetic data, the outliers, the unusual depth at which the specimen was collected all made sense. "I remember getting that photo at night here," says Stiller. "I could already tell that it had extra bones that we don't see in the other two species. I just went home and started looking at the photo in detail."

Morphologically, the ruby seadragon differs significantly from the two other species of seadragon. Seated together in front of a computer monitor, Stiller and Rouse could begin to appreciate some of those differences from the initial photograph alone. The presence of extra bones is important. "You can see how many trunk rings — how many bony rings — they have in their body," says Stiller. "You can tell that from the photo and count up the number of rings, and we saw that there were more rings than in the common seadragon and the leafy seadragon."

There are other subtle morphological differences too: the absence of a head spine; a dorsal spine that points forward in the ruby seadragon and backward in other seadragon species; the particular placement of a third ventral spine; and the color — like a ripe tomato. Before long Stiller was able to tell all three species apart regardless of color, based on the morphological differences alone. If a ruby seadragon is bleached white after decades in a collection, she still can easily discern the differences. In fact, the color of the ruby seadragon is likely an important adaptive development — evolution in action, says Stiller. She suspects the ruby seadragon inhabits the aphotic zone, the perpetually twilit region of the ocean where sunlight struggles to penetrate. This begins to explain its striking coloration. Red light, the longest wavelength in the visible spectrum, is the first color that fails to penetrate to these depths, filtered out by hundreds of feet of seawater. Essentially, in the gloom, even at a depth of about fifty meters, the ruby seadragon is almost invisible to predators. It has no elaborate leafy appendages like the leafy seadragon. It doesn't resemble kelp. It doesn't need to. "An animal that is red will basically be black at depth," says Stiller. "In this case the color could really be a camouflaging adaptation."

Next Stiller and her collaborators embarked on a worldwide search for paratypes — the additional specimens that a taxonomist uses along with the holotype to describe and define a new species. Biologists prefer not to describe a species based on a single specimen. Nothing can be defined with total confidence from a single datum. Since the holotype was captured in relatively deeper waters, Stiller contacted scientific collections and asked to borrow seadragon specimens taken from depths greater than fifty meters. There are two other ruby seadragons, both part of the Australian

National Fish Collection. They've both experienced long shelf lives too — they were trawled from the ocean near Perth, from a depth of seventy-two meters, in May 1956. Then Wilson found the Cottesloe specimen from 1919 in the WAM collection.

"We just pulled all the trays out one day and started going through the dried collection," she says. All together, the dried collection contains almost two hundred seadragon specimens. "They're all laid out in trays. The tray might have eight to ten individuals on it." One by one, she examined every specimen. "Sure enough, we spotted one," says Wilson: a single ruby seadragon specimen with its handwritten label — 1919. "We would have missed the one at the Western Australian Museum," says Stiller, "because that one washed up on the beach. Basically, it fell through my filter of looking for deep specimens."

The ruby sea dragon had stayed hidden for so long simply because the ocean is vast and keeps its secrets hidden. Without a serendipitous moment of discovery, a species that occupies such a limited range, living at depths that render it almost invisible, can remain undiscovered indefinitely; without Stiller's outlier — a strange single point on a graph — it would still be undescribed. In fact Wilson suspects the ruby seadragon washes ashore quite frequently along the remote southern coastline of Australia. "I think a lot of people probably have picked up this species before and just not realized what it was," she says.

In March 2015, a few weeks after the species description was published, a family reported finding a dead ruby seadragon washed up against the rocks at Point Culver in Western Australia. Subsequently, photographs of it appeared online. It was unmistakable: long red body glistening in the sun against the pale gray rock it had been placed on. Recently, says Wilson, a potential collaborator told her he'd found a dead seadragon on the shoreline at Esperance. She asked to see a photograph of it: "It turns out it's another specimen of the ruby seadragon."

In April 2016 Wilson, Stiller, and Rouse embarked on a research expedition to capture images of the ruby seadragon in the wild in its natural environment. Aboard the *Southern Conquest* near Recherche, the team members were all seasick. The ocean swell was twice the predicted size. The boat lurched over range after range of sickening gray peaks. For hours the bow angled into the wet air, then pitched downward — endlessly. The day before, anchored in the lee of a small island in calmer water, team member Daff Phillips had tested an ROV (a remotely controlled underwater vehicle) equipped with a camera, sending it to the same depths the ruby seadragon had been trawled from in 2007. But that was impossible now. Even at fifty meters the swell was too strong to maneuver the ROV.

The next day, with a smaller swell, the team deployed the ROV again. Nothing. Then on the last day of the expedition the ROV moved down through the kelp-dark water to the aphotic zone, and for just a moment it appeared: a solitary ruby seadragon, the color of a ripe tomato. Back in the cramped cabin of the *Southern Conquest*, crowded around the screen, the team erupted in celebration. At the controls of the ROV, Phillips followed the seadragon through the water, skimming above the sandy ocean floor. It led them to another. Then, with its batteries fading, the ROV was pulled to the surface. Minutes later, with the device back at fifty meters and at full power, the seadragon was gone.

"They were the only individuals we saw," says Wilson. Finally the mysterious third species of seadragon had been seen in its natural environment. For just a few seconds in the relative twilight the circle — open since 1919 on Cottesloe Beach — was closed. "We get a bit blasé about the things we think we know," she says. "A common seadragon: it's been around for a long time; it's really distinctive. It's easy to forget that there might be something new."

Museum of Comparative Zoölogy, Cambridge, Mass.

No. 24545.   Paratype "16688"

Oedipus

Thorius pulmonaris Taylor

Cerro San Felipe,

Oaxaca, Oaxaca, Mexico

E. Taylor coll. 18-22, VIII. 38.

Exch. from E.H. Taylor. 17. III. 41.

24545

8

A Century in a Jar:
The *Thorius* Salamanders

In the 1970s, when Jim Hanken was a graduate student, he began to study *Thorius* salamanders, a genus of lungless pygmy salamanders endemic to parts of southern Mexico. At that time the entire genus consisted of just nine known species. They were all minuscule and essentially almost indistinguishable from one another. Today there are twenty-four named species and counting. Hanken, a herpetologist and now the director of the Harvard Museum of Comparative Zoology, discovered fifteen of them.[1]

Thorius arboreus is the smallest of them all. It inhabits wet montane forests in southern Mexico. Its entire life is altitudinous—spent in the mountains in thin, oxygen-poor air. It insinuates its snakelike body into the humid undergrowth. It is superlatively tiny. Slender, with a vibrant bright red band running along its moss green flanks to the tapering end of its bladelike tail, an adult male measures just seventeen millimeters. This makes it, coincidentally, the exact length of my thumbnail. In many other important ways, it closely resembles other *Thorius* species. In some cases even an expert like Hanken—who essentially has now discovered almost all known *Thorius* species—can place specimens of two species side by side and inspect them closely for an hour, nose almost touching them, and still they will not reveal their differences. But with molecular techniques like DNA sequencing, he can tell them apart.

The first *Thorius* species was described in 1869: *Thorius pennatulus*, the Veracruz pygmy salamander. It is almost as small as *T. arboreus*. These days *T. pennatulus* is critically endangered, as are many of the new species Hanken has identified over the past few decades. In the past twenty years *T. pennatulus* has disappeared almost completely. During fieldwork, Hanken just doesn't see it anymore. Although it used to be plentiful, he struggles to find a single specimen.

The species holotype—collected in 1869 in Orizaba by Edward Drinker Cope—is stored at the National Museum of Natural History in Washington, DC. For more than a hundred years after its collection almost no one could distinguish one *Thorius* salamander from another. That didn't stop field biologists from collecting them and storing them in scientific collections and biorepositories. A *Thorius pulmonaris* specimen in the Harvard Museum of Comparative Zoology collection is typical. It looks as if a spindly carrot had sprouted four stubby, half-formed legs. The salamander was collected in the summer of 1938 in the mountains to the north of Oaxaca in southern Mexico. It has spent its time since then in a flask of ethanol.

"These things used to be incredibly abundant, and there are hundreds of thousands of specimens in collections in museums around the world," says Hanken. "They were regarded as so hard to tell apart that a lot of the specimens weren't really identified or were just identified to genus."

In place of specific identification, collections across the world include formalin-filled jars of salamanders labeled simply *Thorius*. To complicate things, sometimes they are identified incorrectly. "People took a guess as to what species they might belong to," says Hanken. Finally, in the 1990s he and his frequent collaborator David Wake—a herpetologist at the University of California, Berkeley— embarked on a quixotic mission: "We basically went through all of the jars in different institutions," says Hanken. They scoured old, established scientific collections in Paris and Berlin and found no *Thorius* holdings at all. Over several years, they visited institutions across the United States. "A lot of what we realized were new species, species that hadn't yet been described, were already sitting in jars. These specimens are pretty small. You could have a hundred or two hundred specimens in a jar."

Sometimes those specimens had sat in preservative, mostly untouched, for fifty years, eighty years, or longer. In London, Hanken found a jar that contained specimens collected in the field in 1902. "They were collected by Hans Gadow, a famous German explorer who had deposited a lot of his material in the Natural History Museum in London," he says. In *Through Southern Mexico: Being an Account of the Travels of a Naturalist*, Gadow described finding *Thorius* species in the high-altitude regions around Orizaba. "They include *Thorius pennatulus*," he wrote, "a tiny newt, less than two inches in length and thinner than a match, with weak limbs and reduced digits."[2]

Unable to identify them properly, Gadow instead sent the salamanders to London. They arrived with numerous other specimens he had accumulated on his expedition: the polymorphic robber frog; the western graceful brown snake; the giant horned lizard. Near Apatzingan he collected the holotype of Bakewell's thread snake. He moved in a wide swath across southern Mexico collecting mostly reptiles and botanical specimens. He netted butterflies too: the pink-spotted cattleheart, the pipevine swallowtail. In his written travel account, Gadow describes *Thorius* salamanders thus: "These little things showed a predilection for living in a proverbially precarious position, namely, 'between the bark and the wood' of decaying pine-trees, amongst the boring-dust of beetles and maggots."

For almost a century, Gadow's specimens — some of them thinner than a match — awaited description at the bottom of a jar. They had been identified, but incorrectly. On a visit to the Natural History Museum in London, Hanken found them. Carefully pouring the alcohol from the jar, he retrieved the specimens from the bottom and inspected them. Hanken suspected there was more than one species of salamander present, but he couldn't know for sure. Later, during fieldwork in Mexico, he collected fresh specimens from the same locations Gadow had visited in 1902. Armed with fresh tissue samples, Hanken used a method called protein electrophoresis to look for differences between the specimens. The technique separates proteins based on their overall electrical charge. All proteins are made up of chains of amino acids, each carrying a small charge. Sometimes two species — though they might look identical in a jar — will have slightly different forms of particular proteins, known as isozymes. The slight differences in charge causes the isozymes to migrate toward an electrical charge at a different rate, allowing evolutionary biologists like Hanken to differentiate them.

Inside the jar were specimens belonging to two undescribed species: *Thorius lunaris*, the crescent-nostriled thorius, and *Thorius spilogaster*, the spotted thorius. Both are found in the vicinity of a scattering of small villages high on the south and southeast flanks of the fluted cone of the Pico de Orizaba, a volcano that is the highest peak in Mexico. They had the longest shelf lives of any species Hanken has described. As Hanken's research illustrates, the *Thorius* genus is a sprawling, diverse group of species, each occupying an exceedingly limited and circumscribed habitat. In some cases several species are sympatric: their ranges overlap. "There are several places in Mexico, several localities, where you can have two species that are sympatric, or even three sympatric species," he says. "In some cases where someone might have put specimens collected in a single place in a single jar, it turned out there was more than one species represented there."

The *Thorius* genus radiated outward, says Hanken, adapting from a common ancestor at least fourteen million years ago, and possibly as early as forty million years ago. The entire genus is distributed across a relatively small geographic area known as the Trans-Mexican Volcanic Belt, and in the Oaxacan highlands of southern Mexico. Some species became arboreal, living, as Gadow had noted, between the bark and the wood. Others remained terrestrial, burrowing into the dark, hidden places beneath the montane forest floor like *T. dubitus*, living exclusively in pine litter. One species, *Thorius magnipes*, lives only in bromeliads. Another species,

Thorius minutissimus, has been found at only a single location: a site near Santo Tomás Teipan, in the Sierra Madre del Sur de Oaxaca, at an elevation of about 2,500 meters. It's possible that it evolved so specifically—to exploit so perfectly the local resources it found there—that, like a key for a lock, it can exist only in that location, in the creases and green folds of the rugged Oaxaca mountains over-looking the Pacific Ocean. It's perfectly suited for the conditions there. Perhaps a few miles to the east, at the same elevation, the microhabitat is imperceptibly different but completely unsuitable.

Sean Rovito of the Laboratorio Nacional de Genómica para la Biodiversidad in Irapuato, Mexico, is another of Hanken's collabora-tors. He has tried to map the order in which *Thorius* species diverged from their ancestor and radiated across the mountains.[3] Rovito estimates that two very similar species—*Thorius grandis* and *Tho-rius schmidti*—diverged very early on, becoming different species about 14 million years ago. Two other outwardly identical species— *Thorius papaloae* and *Thorius magdougalli*—diverged an estimated 5.7 million years ago. For context, humans diverged from chimpan-zees about 6.5 million years ago. In other words, humans and chim-panzees share about the same information genetically as *T. papaloae* and *T. magdougalli,* even though, when placed next to one another, the two salamanders are essentially indistinguishable. And it's not just their small size that makes identification difficult, Hanken says: "Even if you were to blow them up to a larger size, they just don't differ than much. It's more subtle."

Once Hanken has identified a new species, he often finds ana-tomical features he can use to differentiate it from others. There will be a slight but measurable difference in body size or external color-ation or in their osteology—the presence or absence of a particular bone. To determine osteological differences, he uses histology: "If you make a cleared and stained preparation—when you stain the bones one color and the cartilage another color, and you make the rest of the body transparent—you have the skeleton displayed in front of you," he says.

Knowing whether there are twenty-five—or thirty, or fifty, or however many—*Thorius* species distributed across southern Mexico instead of just a handful is important. "It has great implications for conservation strategies," says Hanken. Conservationists can't protect a species until they know it exists. "If it turns out that what you thought was a single widespread species actually consists of twenty-five species, each with a very small range, then your whole strategy for conserving that genetic heritage and biodiversity is very different."

Either way, conservationists will struggle to save them all. The damage is already done. For many species, their habitats are disappearing faster than they can be identified—cleared for agriculture or urban development, destroyed by illegal logging, or affected by climate change. Even so, Hanken is still naming new *Thorius* species. In November 2016 he and his coworkers described three new species of *Thorius* salamanders, all from the mountainous parts of southeastern Oaxaca.[4] By closely inspecting specimens that look the same, he is continuing to realize that they are not. The novel species were all collected from a region previously delimited to one species, *T. minutissimus*. Hanken collected the holotype for the pine-dwelling minute salamander (*T. pinicola*) near the village of San Miguel Suchixtepec in July 1976, from a scrubby region of pine-oak forest and red dirt. *Pinicola* comes from the Latin *pinus*, for pine, and *cola*, meaning "inhabitant of." Another species, the long-tailed minute salamander (*T. longicaudus*) had been collected from Sola de Vega forty-five years earlier, in November 1971—long and thin-bodied in photos, like a discarded shoelace. A coppery stripe runs along its wet back. When Hanken visited the type locality in 1974, he saw hundreds of individuals that had crammed themselves into the cracks in the dirt beneath the roadbed—ten or twelve in each crack. By 2014, when his coauthor Sean Rovito visited the location again, he couldn't find even one. The area had been changed by logging: the road had been widened, and the salamanders were gone. A living specimen was last seen in 1998. Specimens of the third new species, the heroic minute salamander (*T. tlaxiacus*), date back to 1976. Finally Hanken and others can begin to understand the species' role in the complex biodiversity of the Oaxacan ecozone, even though they might no longer be there to perform it.

There are others: "Right now I know of at least eight more un-described species of *Thorius* sitting in jars, for which we have data," says Hanken matter-of-factly. Elsewhere around the world—in basements and on shelves—there are even more unknown *Thorius* species, suspended in flasks like burnt matches, waiting to be named.

9

From a Green Bowl:
The Overlooked Squeaker Frog
(*Arthroleptis kutogundua*)

A few years ago, David Blackburn found himself standing in front of a large glass jar at Harvard University. It sat on a shelf in the Museum of Comparative Zoology, on top of a compactor cabinet housing the herpetology collection. In the jar, he recalls, were a lot of frogs. "Fifty or sixty medium- to large-size frogs," says Blackburn, who is now a curator of amphibians and reptiles at the Florida Museum of Natural History. "They're all brown. They're all contorted."

At some point in the past, the entire jar of frogs—every last brown and contorted specimen—was identified as a single species: *Arthroleptis adolfifriederici*. The label is on the jar. Also known as the Rugege Forest squeaker frog, *A. adolfifriederici* is a small and relatively common frog found throughout Central and East African countries. It was first described in 1911. On those grounds alone the contents of the jar interested Blackburn. "I work on African frogs," he says. "I happen to work on this group—*Arthroleptis*."

The specimens were stored in ethanol, which had, Blackburn recalls, become "a tad yellowish" with age. He estimates that at least thirty years—and perhaps forty or fifty years or longer still—had passed since anyone had looked critically at the frogs in the jar.

"One of the things I realized as a graduate student," says Blackburn, "even from looking at the jar from outside, was that there were more than one species in this jar. There was a lot of variation in there." Carefully he lifted the jar from the shelf, removed the lid, and began to sort through its contents. "For me, the way of doing this was just simply sitting with a tray and, one by one, pulling them out and trying to sort them into piles," he says.

"I'm going through specimen by specimen," says Blackburn. "All they have is a number. There's nothing else written on the tag. It's just a number." As he pulled the frogs from the jar with long-handled forceps, one specimen seemed different from all the others. "I pulled it out and put it in its own pile," he says.

Before long the jar was empty. Arranged on the tray in front of Blackburn were several piles of frogs, which he'd sorted into recognizable groups. The one strange and unidentifiable frog remained by itself, a single specimen, #16952. Blackburn named and described the frog in 2012, in a *Journal of Herpetology* paper.[1] He gave it the scientific name *Arthroleptis kutogundua*. Its common name is the overlooked squeaker frog. It had waited more than eighty years to be described. "Basically, this is a frog described from a single specimen," says Blackburn.

Arthroleptis is a widespread genus of about fifty frog species found across tropical sub-Saharan Africa—also know as squeaker or screeching frogs for their characteristic high-pitched calls. They are typically small, terrestrial species, often only a couple of centimeters long. "As far as we know," says Blackburn, "all these frogs lack a tadpole. These are direct developers that hatch directly out of an egg."

Other species include the Freetown long-fingered frog (*Arthroleptis aureoli*), the Mount Nimba screeching frog (*Arthroleptis nimbaensis*), and the cave squeaker (*Arthroleptis troglodytes*). Across the continent, many *Arthroleptis* species are critically endangered—their habitats are being destroyed, and the frogs with them.

A herpetologist inspecting an eighty-year-old amphibian specimen has to develop a set of critical skills that, say, a lepidopterist studying a well-preserved butterfly doesn't need: the ability to extrapolate. The specimen no longer looks the way it did in life. In other words, faced with the part, Blackburn must attempt to visualize the whole. This is especially true for frogs. It's different for reptiles. Reptile systematists often rely on scalational differences — distinct features in the color patterns, positions, and numbers of scales on a specimen. A unique snake species might have an extra scale in a specific location, or a series of differently shaped scales. In 2015 University of Wolverhampton herpetologist Mark O'Shea named *Toxicocalamus ernstmayri*, a venomous worm-eating snake from the Star Mountains of New Guinea. The holotype—it is the only known specimen—also is part of the herpetology collection at the Museum of Comparative Zoology at Harvard, where it had been stored for almost fifty years.[2]

Early field expeditions to places like New Guinea or Tanzania typically didn't even include a herpetologist, says O'Shea. When collecting specimens, their main interests were entomology, ornithology, mammalogy, and botany. "Herps were a by-catch that one of these people, or the expedition doctor, took charge of," he says. The expedition team then sent herpetological specimens back to their home institutions, expecting someone there to examine them. "But, of course, either they didn't for the same reason—lack of expertise—or it was many years later when the collectors were dead and unable to provide more data," he says. "Even the seven Archbold expeditions of the 1930s and 1950s to New Guinea, which made large herp collections, didn't have a herpetologist on the team."

O'Shea had noticed the snake, a bronze-colored three-foot-long specimen collected in 1969, on the shelf at Harvard. He used the number and arrangement of scales to identify and describe it. The specimen was reduced to a map of individual scales: 203 ventral scales, 29 subcaudal scales, 6 supralabial scales, and so on. For this reason it is easier to recognize undescribed species of reptiles in old collections than frogs. Those differences remain. But after decades in preservative a frog specimen slowly becomes a strange, pale ghost of itself. There are no scales to count. The skin markings on the frog that Blackburn removed from the jar—bold and iridescent in life—had become muted and indistinct. In life, frog species have been described by using their vocalizations as a means of identification. In 2014, researchers named *Rana kauffeldi*, or the Atlantic leopard frog, a new species, living on Staten Island in New York City.[3] Its discovery depended on its distinctive abrupt call, which separated it from its closest relatives. In the case of *A. kutogundua* there are no sounds left to analyze.

The holotype is dark like a lump of waterlogged tree bark. One of its long hind legs is raised and bent at the knee. The other leg hangs down limply, as if frozen mid-jig. Its stubby front legs are by its sides. It has a broad, shovel-shaped head. There is a small incision on its abdomen—its intestines are gone. But it still looks like a frog. Small, with long, elegant toes—but a frog. "In many ways this is an unremarkable animal," says Blackburn. "It's remarkable among the squeaker frogs because of its size—very few of them are this large—but this isn't even that large for a frog. It's a small frog, about four centimeters long."

The holotype was collected on March 19, 1930, by Arthur Loveridge, a British herpetologist who was curator of amphibians and reptiles at the Harvard Museum of Comparative Zoology. Loveridge was a hardy and intrepid explorer. During World War I he had served in the East African Mounted Rifles, a unit made up of British settlers who were living in British East Africa when the war began. Later, as a field biologist, he made numerous collecting expeditions to Africa, spending months at a time in dangerous and unmapped terrain. Loveridge found the specimen within the forested crater of an extinct volcano in remote southwestern Tanzania. Ngozi crater is steep-walled like a cauldron. At the bottom there is a lake.

"Camped on the narrow lip of the crater from March 18–20th, 1930," wrote Loveridge in "Reports on the Scientific Results of an Expedition to the Southwestern Highlands of Tanganyika Territory."[4] When he found the frog, Loveridge was halfway through an eight-month expedition through the southern highlands. He'd arrived in Dar es Salaam in early November 1929. By March it was the rainy season. It rained constantly, and clouds hung on the volcano.

On March 18, 1930, Loveridge arrived at Ngozi: "It was 3 pm on the day of our arrival before we got the lip of the crater sufficiently levelled off to be able to pitch, or rather sling, two tents between cables affixed to trees growing up from the precipitous sides of the crater."

On the morning of the second day, accompanied by three native assistants, Loveridge descended into the crater to collect specimens. They scrambled down the steep green walls of the cauldron in the rain, over saturated ground. They combed the leaf litter and turned over logs. In almost five hours, they collected only forty frogs.

"We took a single specimen of *Arthroleptis adolfifriederici*," wrote Loveridge. He found it hopping on the sodden forest floor. Eventually, along with all the other specimens from the expedition, the frog was sent to Harvard University in Cambridge, Massachusetts. There it became specimen 16952 in the Museum of Comparative Zoology herpetology collection.

How did Blackburn identify the single *A. kutogundua* specimen in a jar filled with similar old frogs? This question occupies the center of species discovery. What allowed him to see the part and somehow visualize the whole? Even now, Blackburn says he doesn't know. He can't explain the intuition that led him to place the specimen in a pile of its own. A pile of one. What details separated one unremarkable frog—brown and shriveled after eighty years in alcohol—from the rest? "It was one of those things where I feel like, by gestalt, I knew it was different," he tells me, "but I couldn't have articulated it to you."

Perhaps Blackburn simply has a mind awake to the mysteries that remain undiscovered in the collections. Across his career, he has inspected tens of thousands of frogs—either in institutions or in the field, on expeditions to Angola, Cameroon, Malawi, Nigeria, and Uganda. Taxonomically, he says, the frog just didn't belong to any of the *Arthroleptis* species he'd seen before. He knew the specimen was different without having to subject it to the multitude of careful anatomical measurements he later included in its formal description. He carefully converted the specimen into data—a constellation of measurements including eye diameter, thigh length, and the length of its digits. It has a small tympanum, he says, and long, slender digits. But the data merely confirmed in a more irrefutable and empirical sense what Blackburn's intuition had already told him—that it was new.

In the local Kiswahili language, *kutogundua* means "to not discover."

Species that might seem unremarkable—like the overlooked squeaker frog—can have an enormous and far-reaching environmental impact. In fact, unseen in the undergrowth, they can even drive an entire ecosystem.

"A great example of that is some recent work done on salamanders of the genus *Ensatina* that are found in California," says Blackburn. "It's a ring species—something that basically grows around a geographic barrier and divides and slowly differentiates so that by the time it comes around the barrier, in this case the Central Valley of California, it no longer recognizes itself."

Like *Arthroleptis kutogundua*, *Ensatina* salamanders are small and inconsequential-looking. Individuals are about two centimeters long. "They're found everywhere in the redwood forests," says Blackburn. "These are inconspicuous small animals, but some recent work has shown that they're actually incredibly important for moving carbon through the environment because there's a lot of them and they eat a lot of terrestrial invertebrates. When they die and when they defecate, they're cycling carbon back into the soil. So, even though they seem small and inconspicuous, they can be really important to these ecosystems like redwood forests."

Hidden in the damp leaf litter of sub-Saharan rainforests, *Arthroleptis kutogundua* might play as vital an ecological role as *Ensatina* salamanders. It could be a living engine, driving enormous unseen processes—if it still exists. In recent years field biologists and conservationists visiting the area haven't reported seeing anything like it there.

"I had a lot of colleagues in East Africa, who do work there," says Blackburn. "I asked around. I even floated part of the description by some colleagues just to say, 'I know you've been working in southern Tanzania, have you ever seen anything like this?' They all said no."

For the past several decades, says Blackburn, biologists have watched as many frog species worldwide rapidly dwindle in numbers. "There's a frog species in Cameroon, in the mountains, that was undescribed." Finally, a couple of years ago, some of his colleagues described it. Ten years earlier, the species had been widespread, found everywhere in the mountains. "It was this species that you'd find and you'd say: 'All right, we've seen so many of these things, don't even bother picking it up,'" says Blackburn.

But that has changed, and quickly. "What used to be one of the most common things," he says, "the thing that nobody even collected sometimes because it was just so common, now you don't even see it. The question is, What's happening?"

By some estimates, almost a third of frog species worldwide are threatened with extinction. Among the many complex factors involved are climate change, habitat loss, and the introduction of invasive species. In many regions amphibian populations have seen dramatic declines from chytridiomycosis — an infectious disease caused by the fungus *Batrachochytrium dendrobatidis*. Some species have disappeared altogether. Joseph Mendelson, herpetologist and director of research at the Atlanta Zoo, has seen the decline firsthand. "We've lost a good number — an unknowable number — of frogs," Mendelson tells me. "We're never going to actually know what we lost in the last twenty years or so in Central America, which is where I do most of my work."[5]

In some cases entire groups of once common and diverse amphibian families have almost disappeared. An example is the harlequin toads — genus *Atelopus* — endemic to the Andes. "There were over 225 species of those," says Mendelson. "I think seven, maybe ten, of them are currently findable. The rest of them are just gone. And they've been gone for decades. They're almost all certainly extinct as a result of this chytrid fungus."[6]

Hoping to find additional specimens of *A. kutogundua* for his description, Blackburn asked his colleagues around the world to search their collections. But there weren't any. Then, a year after the description was published, University of Roehampton herpetologist Simon Loader suddenly found another one. In March 2013 Loader was visiting the Museum für Naturkunde in Berlin when

Frank Tillack—the herpetology collection manager there—asked him to sort through some specimens stored in an old box of material from East Africa.

The frogs in the box were unidentified. Loader picked through them: "When I saw the specimen, and the locality of southern Tanzania, I was aware that there haven't been many large arthroleptids found there," says Loader. "Once I'd ruled out its belonging to the most likely common species of *Arthroleptis stenodactylus*, I knew it was something interesting and was likely to be Dave's new species."

Loader was right. Tillack sends me a single photograph of the specimen, laid out on its belly next to a ruler. Totally devoid of pigment, its skin is transparent—a strange pale yellow—like a plastic toy frog that hasn't been painted yet. It was collected on October 27, 1899, by Friedrich Fülleborn. With only two known specimens in existence, the natural history of the frog remains a complete mystery.

"We have so little data about it that we can't actually place it into the Tree of Life," says Blackburn. It stands alone—known, yet not quite known at all. Regardless, Blackburn is hopeful that *Arthroleptis kutogundua* can still be found in the rugged southern highlands of Tanzania. Perhaps it clings on, a small but shrinking population on the steep forested walls of Ngozi crater. "My hope is that it's still there somewhere," he says. Perhaps *A. kutogundua* has never ventured beyond the green bowl formed by the crater. Maybe, over millions of years, it slowly became a species there, then lived there, and finally disappeared there too—its entire history spent in the crater. It remains a mystery.

"Here's a species of frog from a region that's relatively well explored for which we still have only two specimens," says Blackburn. "The two times it's been seen are 1899 and 1930. We have a lot better access to these regions now than we ever did in the 1890s or 1930s. Certainly there are biologists who regularly visit the area, and we've never turned this thing up again."

Blackburn pauses for a moment. "In a sense," he says quietly, "it's rarer than a *T. rex*."

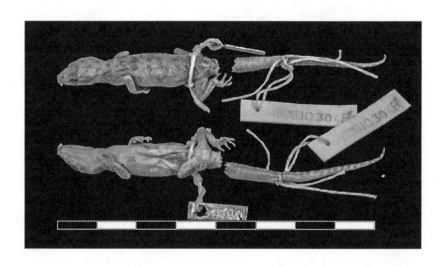

A Body and a Disembodied Tail: Smith's Hidden Gecko (*Cyrtodactylus celatus*)

Cyrtodactylus celatus is a pale yellow gecko patterned with random irregular brown splotches—faded now but still visible. Its skin is wrinkled and its eyes are dull gray like pencil lead. The specimen, an adult female, has long, tapered toes not at all like the flattened adhesive toe pads found in many other gecko species. Its tail is detached from the rest of its body. Now its abbreviated body ends in a ragged stump where the tail broke away. But everywhere the body goes, the tail goes too. A length of string is knotted around the middle of the tail with a label attached so the two parts of the specimen can't become separated forever. Along its length, the gecko is stippled with tubercles—warty little outgrowths—one of the unusual features, says Hinrich Kaiser, that allowed him to identify it as a new species.

Kaiser discovered the gecko while working at the Natural History Museum in London. It sits in a small jar filled with alcohol at the Darwin Centre in South Kensington, an impressive eight-story modernist structure shaped like a cocoon, which seems to float supernaturally within a tall glass-sided atrium. Kaiser was in the herpetology collection there, unraveling snakes. It's slow and painstaking work. One after another, he reached into a jar of yellowing alcohol and removed a snake. Some jars contained two or three or more snakes, tangled together like strings of Christmas lights-their long, pale bodies spiraled in elegant helices against the glass. He pulled them out, uncoiled them like stiffened electrical cords, and arranged them on a tray in front of him.

Kaiser, a herpetologist at Victor Valley College in Victorville, California, eighty miles or so northeast of downtown Los Angeles, was in London to study snakes from Timor, where he conducts most of his fieldwork. He'd come a long way to inspect the snakes, and his collaborators had traveled from Germany to meet him there.

"When you're at the British Museum, you don't just look at one thing," he says. Once he'd finished uncoiling and measuring Timorese snakes, Kaiser turned and asked Patrick Campbell, the senior curator of reptiles, "What else do you have from this area that might be relevant to us?" Everywhere they looked, shelves were lined with glass jars, together containing thousands of amphibian and reptile specimens. "This is where you stumble on new things," says Kaiser.

A label on the gecko's jar he was shown reads *Gymnodactylus marmoratus*. Inside the jar, alongside the wrinkled body of the gecko and its floating disembodied tail, was another label, handwritten in pencil in the 1950s, which added to the confusion: *Cyrtodactylus wetariensis*. But the specimen is neither.

In 1924 Malcolm Smith and his wife spent two months in the Malay archipelago collecting herpetological and botanical material. As the weeks passed they hopped from island to island, obtaining specimens as they went. In the rugged green mountains of East Java, Smith collected red stream frogs. In Indonesia, on Sulawesi—he called the island by its earlier name South Celebes—Smith ascended Mount Bonthain, an extinct volcano, struggling upward through relentless rain and fog into the thin air. "From the branches of all the trees at the higher levels, long festoons of pale grey or white lichens were hanging, giving the scene a most wintry aspect," he wrote in an account of the journey.[1]

Smith was forty-eight then. Decades earlier, in 1902, he'd left his post as a surgeon at Charing Cross Hospital in London for Bangkok, where he became the medical officer for the British Legation. He stayed in Thailand more than twenty-five years and became the personal physician to the Court of Siam.

Smith was also an amateur herpetologist. There's a photo of him as an older man in a 1956 *Life* magazine article on amateur naturalists: thigh-deep in dark water in an English pond, dapper in a tweed suit and a hat, with a frog net in his hand, dripping water.[2] Born in 1875, Smith was a true Victorian scientist. His early articles, published in the *Journal of the Natural History Society of Siam*, reflect a brand of scientific method no longer practiced: the scientist and his dutiful servants hacking through the unbroken jungle, followed by afternoon tea with the Major on the manicured lawn of the sports club. In a September 1913 report titled "Large Banded Krait," Smith wrote, "An unusually large specimen of the Banded Krait (*Bungarus fasciatus*) was recently killed in the compound of the Bangkok Nursing Home. It was trodden upon by the house coolie when going out to fetch water after dark, and was promptly dispatched by the remainder of the staff who came to his assistance."[3]

Or this, from a 1914 article on the common rat snake: "The largest specimen that I have seen was killed one afternoon on the lawn of the Sports Club, where it had taken up its position beneath a chair, in broad daylight and with people about. It measured 2110 mm (6.11 in.), and had just eaten three large toads, a fact which probably accounted for its incautious behaviour on this occasion."[4]

You can almost hear him harrumphing as you read his accounts. Eventually, through his unstoppable persistence, Smith became an authority. The amateur became the expert. In 1926 he published a seminal monograph on sea snakes, in which he described numerous new species collected from across Asia.[5]

Timor is an island north of Australia in the north Timor Sea, in the southern arc of the Lesser Sunda Islands. Remote and undeveloped, the island is still a difficult place to reach. The land is a mixture of coastal scrub, rainforest, and rice paddies—almost all on a difficult slope. The island is divided in two: West Timor is part of Indonesia;

East Timor became a sovereign country in 2002. In 1520 Portugal claimed the island as part of its expanding territories. The Dutch arrived in 1640, occupying the western half of the island and forcing the Portuguese to the east. One way or another, the island has seen unrest and turmoil for centuries. There is blood in the soil: first came attempts at European colonization; then came Timorese resistance, followed by Japanese occupation, Indonesian invasion, and guerrilla incursion. For decades, political conflict has kept field biologists away. This is another aspect of the taxonomic impediment: to properly know the fauna of a place like Timor, biologists first have to get there. Once there, they might encounter demanding and unpredictable circumstances that hamper the work—conflict, corruption, the lack of infrastructure. There are reasons that stable regions are better surveyed than difficult places like Timor.[6]

When Smith arrived at Kupang, the largest city in West Timor, at the end of February 1924, he was unimpressed. "From a herpetological point of view," he wrote, "Timor is one of the most disappointing places that one can visit. Considering its size and position in the Tropics, there is probably no other island so barren in reptilian and batrachian life."

Nevertheless, he did find and collect life there. The gecko came from Djamplong, thirty-five miles northeast from Kupang. When he trapped it, in a wooded area braided with numerous small streams, its tail detached—a defense mechanism called autotomy. Afterward, along with the other specimens he collected, Smith sent the gecko and its tail to the Natural History Museum, where it has remained since, now overseen by Patrick Campbell in the Darwin Centre.

Ninety years later, Hinrich Kaiser found the gecko and named it *Cyrtodactylus celatus*, from the Latin *celatus*, meaning "hidden, covered, or concealed."

The story of the gecko is common enough. The specimen stayed hidden for so long—and in a cataloged and well-used collection, too—because the taxonomy of bent-toed geckos from Timor and nearby is a jumbled mess, filled with errors and inaccuracies. "In the early days of collecting there was, first of all, a lot of collecting done that was perhaps not quite in really confirmed localities," says Kaiser.[7]

When field biologists recorded the type locality as Timor, the specimens might instead have come from an array of nearby islands—from tiny outcrops like Rote, or Sawu, or Alor, or from places so small that no one has even bothered to name them. Adding to the confusion, when specimens were stored in biorepositories they were misidentified—placed in jars with similar-looking species—or mislabeled, identified only to genus or sometimes to the wrong genus altogether. This is a problem worldwide, in all collections. In a November 2015 paper in *Current Biology*, researchers from Oxford University and the Royal Botanic Garden Edinburgh estimated that as many as 50 percent of natural history specimens are wrongly named. Typically, says Kaiser, in the United States, natural history collections grew because a curator with an interest in a particular taxon went on expeditions and collected specimens by hand. Not so in Europe.

"Every wealthy merchant and every pharmacist and every aristocrat had his own little cabinet of curiosities," Kaiser says. "They just assembled anything they could. That, then, was the background upon which the early European collections were founded." This was particularly true for collections in German-speaking and Dutch-speaking countries, says Kaiser, and by extension the countries they colonized—like Timor.

"In England you had the king," he explains. "In France you had the king and then an emperor. Elsewhere in Europe you had essentially all these counts and little kings, and everybody wanted to do their own thing. You literally had dozens upon dozens of these little collections. Nobody really knew anything about the fauna they were collecting. These are some of the things we're finding and looking through now."

These cabinets of curiosities were the first collections. Some of them gradually became part of larger collections, like the Natural History Museum collection in London. But what if half of its specimens have been given the wrong names?

Since East Timor was formed in 2002, Kaiser has made several expeditions there. It's still a difficult place to work. "You need to turn over a lot of things, and walk a long way, before you find anything," he says. "Some nights with a full moon, it's terrible." But on humid moonless nights—wading through the rice paddies and navigating coffee plantations in the oil-black darkness—they did better: arboreal pit vipers, wormlike blind snakes, lots of foam-nest tree frogs. Rice paddy frogs. Monitor lizards. Wedge skinks. Kaiser collected more *Cyrtodactylus* geckos, too. Most of them have become part of the Smithsonian Institution's collection. They were similar, he says, but not identical, to the single specimen Smith collected to the west, near Djamplong, in 1924, which remains the only known specimen of a species that might no longer exist. On a recent expedition, Kaiser collected specimens of about twelve new species of bent-toed geckos. "They are totally new," he says. "There's no museum material whatever."

To confirm that the species he collected are unknown, Kaiser will compare them with type specimens already in scientific collections—a painstaking process that can take years to complete. Lining up countless century-old gecko specimens, Kaiser will count and map their scales and take X-rays to identify subtle osteological features—work that can still be performed on a ninety-year-old specimen. But it's a slow process. "It's a huge amount of work to see all the type material," he says. Kaiser has crisscrossed Europe—London, Vienna, Leiden—traveling from one collection to the next for five years, removing holotypes from locked cabinets. He has carefully assessed thousands of geckos. In doing so he's uncovered even more unidentified species. By now he's lost count of how many. When I ask him for a total, he starts to count quietly to himself, stops, and starts again. Finally he gives up.

"It's dozens," he says. He begins to list them: several species of frog; five skinks; geckos collected in 1920; a few snakes, he says— one described from a single specimen collected in Java in 1937.[8] "We have some from the Senckenberg Museum in Frankfurt," he says. "There are some from the Museum für Naturkunde in Berlin. We have some from Vienna—the Natural History Museum. I forget if we have some from Basel."

Everywhere he went—in decades-old jars and vials—Kaiser has discovered unnamed *Cyrtodactylus* species. "We found a new species from Bali, three new species from Lombok, a new species from almost every one of these little islands," he says. "When you add all those up—and this is just for the genus *Cyrtodactylus*—you end up with a grand total of twelve new ones that we recently collected ourselves, plus twenty-one that have been sitting in museums."

He plans to name several of the geckos for the forgotten heroes of the Timorese resistance, but it will take him years to describe them all. Other unnamed species wait in the Darwin Centre in London and in collections everywhere. Kaiser will continue to visit Timor, too. There, on black, moonless nights, he will search the coastal woodlands and the thorn scrub, following the beam of his headlamp as it bounces around in the dark.

The Invertebrates

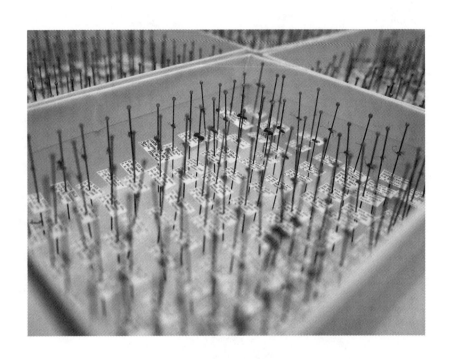

11

Treasure in the By-Catch: The Gall Wasps (*Cynipoidea* species)

In September 2015 Matthew Buffington was behind the wheel of an air-conditioned Chevrolet cargo van headed east. Outside, the heat was otherworldly. The buttes and mesas of southern Arizona, red and orange, unspooled on both sides of the van. Buffington, an expert on parasitic cynipoid wasps, is an entomologist in the Systematic Entomology Lab at the United States Department of Agriculture. Normally he can be found in his lab space at the National Museum of Natural History in Washington, DC. But in 2015, as the summer drew to a close, he took a road trip.

"I drove our lab van from Washington, DC, to the California Academy and hit major collections along the way, picking up alcohol samples that we can't ship anymore," he says. "I was just moving stuff in a vehicle. Part of my motive was to go to collections I'd never seen before."

Buffington drove more than seven thousand miles in two weeks: west in his van from Washington, DC, across the country to San Francisco, south almost to the Mexican border, then back to Washington, DC, again in an enormous country-size loop. Stowed in the back of his van were approximately 3.5 million insect specimens, wasps and beetles mostly, from Madagascar and other difficult-to-reach places. Buffington had picked them up a few days earlier from the California Academy of Sciences in San Francisco. Ultimately, he says, it will take him several years to assess them, let alone describe and characterize them all.

But as he drove on—eastward through Tucson to College Station, Texas, and later to Mississippi—Buffington thought about Logan, Utah. Like a neat and orderly grid laid across the pan-flat valley, Logan is a small town in northern Utah. It sits near the southern border of Idaho, hemmed in to the east by the northern limits of the Wasatch range—a scrubby band of peaks that rises abruptly from the valley floor and stretches south for almost two hundred miles. On the eastern side of Logan, with an uninterrupted view of the range, is the campus of Utah State University. The USDA bee research facility is there too, and it houses one of the largest collections of bee specimens in the world. "It's unbelievably huge," says Buffington. He went to Logan to collect alcohol-preserved samples that a colleague had recently collected in Bolivia and Peru and the desert southwest. "As it turns out, when I got there he was more or less not ready for me," he says. "The freezer was a mess. Things were disorganized."

After a brief attempt to sort through a mismanaged freezer, Buffington decided to take a look at the pinned bee and wasp specimens—each specimen mounted with an entomological pin.

In 2015, just before his road trip, Buffington and his colleagues had published an exhaustive 177-page article in the journal *ZooKeys* on afrotropical cynipid wasps—a large, diverse group of small black-winged insects.[1] For that work they relied heavily on specimens from collections around the world: Berlin, Brussels, Budapest, Cape Town, Hamburg, London, Nairobi, Paris, San Francisco, Stockholm, Washington, DC, and elsewhere. They came from some of the largest and oldest collections in the world, not from unlikely places like Logan, Utah. Despite borrowing hundreds of specimens from the collections, several previously described afrotropical cynipid species were missing. Though he knew they existed, Buffington couldn't include them in the monograph. They are simply so rare that he couldn't find representative museum specimens.

Back in Logan, "I go into the cabinets," says Buffington. "I find the section I'm interested in, and I start pulling out drawers." Immediately he was surprised by the number of drawers dedicated to cynipid wasps. Most decent-size insect collections, he says, might have a couple of drawers filled with specimens. "That's about five hundred to two thousand specimens," says Buffington. "For our purposes, that's meager. That's very small. It's almost like an afterthought. Here at Logan there were something like twenty drawers."

Buffington worked his way through the specimens, skimming through different genera like a researcher browsing books on a library shelf. Then he arrived at the unsorted material: "It's this shitload of material from all over sub-Saharan Africa," he says. "Mostly Mozambique, South Africa, Angola—places that just haven't been collected recently. The collectors, they're all folks who worked at the Bee Lab; they go all over the world collecting their bees, and in this case it seems like they kept everything. Everything that was non-bee insect, instead of discarding it or ignoring it, they actually brought it back."

Mostly the specimens were collected in the 1950s and 1960s. There are older specimens too—the oldest was collected in 1923.

Cynipid wasps are members of the order Hymenoptera, which includes bees, wasps, ants, and sawflies. They're tiny, and they don't at all resemble the archetypal black-and-yellow-striped insects that people imagine when they think of a wasp. Instead, they are small black flies-most of them only a couple of millimeters long. Even the largest species measure only about eight millimeters. They don't sting. Like all Hymenoptera, they have a very specific anatomy that they all share regardless of their differences: a head with compound eyes; a mesosoma, which includes the thorax and wings and the first segment of the abdomen; and finally the metasoma—the large posterior segments of the abdomen.

The group is divided into five families. The unsorted specimens in Logan—thousands of them—had been identified to family level only.[2]

Otherwise known as gall wasps, cynipids are parasitic species.[3] The female wasp lays her eggs with an ovipositor on an actively growing part of a plant. By a process that is still not fully understood, a gall then forms on the plant, creating a highly distinctive plant structure. The gall is made of the plant's tissue but is controlled by the wasp. Essentially, the wasp harnesses the plant's machinery to form a protective structure. The vulnerable wasp larva develops safely inside an internal chamber, using the gall and its contents as shelter and food.

Galls look different depending on the species responsible for them. Made by a horned oak gall wasp (*Callirhytis cornigera*), a woody twig gall growing on an oak twig looks like a smooth, rounded nut, strangely wrapped around the wood. The parthenogenetic gall wasp (*Andricus foecundatrix*) lays a single egg in the leaf bud of the pedunculate oak (*Quercus robur*). On the branch, surrounded by normal leaves, that one bud begins to distort, driven to reshape itself by the insect growing within it, developing into a green fluted structure that resembles a miniaturized artichoke. The gall falls to the ground with the leaves in autumn, and the larvae overwinter inside, beneath the tree. In the spring, mature wasps emerge. The variety in galls is almost endless. The common spangle gall wasp (*Neuroterus quercusbaccarum*) covers the underside of oak leaves with small purple discs like splatters of paint. The *Diplolepis rosae* gall wasp infects the field rose with a gall that resembles a mossy green pincushion. Inside, the woody core of the gall contains several chambers, each with a single larva. Other galls look like pieces of yellow coral, or brown peas striped with a cream-colored whorl, or brown droplets hanging from a leaf. On the scrub oak, the beaked twig gall wasp (*Disholcaspis plumbella*) creates bright red, almost spherical galls covered with yellow dots the color of egg yolk, with a single protruding point at the apex.

Not all cynipids are gall-forming wasps. In fact, Buffington is particularly interested in species within the larger family—and there are many of them—that act as parasites to other insects, laying their eggs in their larvae and using them as unwitting hosts. In some cases parasitic wasps have been intentionally introduced into agricultural ecosystems, where they protect valuable crops by killing the pests that destroy them. Buffington sees those species as powerful weapons against agricultural pests.

Then there are the hyperparasitoid cynipids — they parasitize the parasites. Armed with a long, flexible ovipositor, the female deposits her eggs within the internal chamber of an occupied gall. When the eggs hatch, the larvae feed on the gall-forming larvae already inside.

The specimens in Logan had been point mounted: a pin was driven through the base of a long isosceles triangle cut from cardboard, and the specimen was glued by one side of its body to the point of the triangle farthest from the pin. The method provides a secure hold that allows entomologists to see all sides of the specimen. The wasps are in boxes — a forest of pins. Here and there containers hold galls too, like little rounded nuts. The wasps are far too small for one to appreciate their anatomical subtleties by eye, but under a microscope a cynipid wasp is beautiful: black and streamlined like a miniature fighter jet, its shortened, winged thorax rises steeply behind the head, meeting a glossy, plump multisegmented abdomen. The metasoma shines as if coated with varnish. Some species are narrow-waisted — the metasoma attached to an elongated mesosoma via a long, thin stalklike petiole. On each side of its head, compound eyes glint like chrome mesh. When Buffington looked at the label data beneath each specimen, he was astonished. "I'm expecting Utah, Idaho, Wyoming," he says; "I see stuff like Transvaal, Western Cape."

And the moment he saw them — tiny black specks glued to the points of cardboard triangles — Buffington understood that he was looking down at a plurality of new species. "When I look in a drawer, one of the first things I assess is How full is it? That tells me how much of it has been sorted or not. If there's lots of space, it suggests they don't really have that much backlog. It means most of it's already been identified."

Buffington immediately noticed something: "These drawers were packed." There was incredible variety of species, with some of the largest gall wasps in the family alongside species almost too small to see. "That suggests this collection is very broad — it hasn't just focused on one thing. It's not all just from one trap or one sweep net. It's a big mixture." In fact, the collection is compendious. Even the rare species he'd excluded from his 2015 paper were there. "This is an entire genus that I didn't get into my monograph because we can never find any specimens that represent it, and here they are in Utah, collected in western Cape Province in the 1940s. I would have never figured that, you know?"

The Department of Entomology at the National Museum of Natural History is home to more than thirty-five million insect specimens. Inevitably, some of the collection—perhaps even most of it by some estimates—remains unknown. This is true for all large insect collections. "When it comes to numbers, we only have names for about 10 percent of the actual insect diversity," says Buffington. "Chances are there's shit in museums that's undescribed, just from a numbers game alone."

And Buffington likes those odds. "When I go to the British Museum, I'm not too interested in what their identified stuff is. Usually it's wrong. The labeling is somewhat questionable. The locality data, a label that says 'Argentina, 1990'—scientifically, that's almost a useless specimen."

Instead he heads for the drawers and cabinets holding specimens that have no names at all. "The real fieldwork, the moment of discovery, all takes place in what we call the unsorted material," he says. "Most of us who cruise around looking at collections, that's where we're spending our time, because we know, we actually know going into it, there are undescribed things there." Essentially, the Logan collection is an untapped reservoir of unsorted Cynipoidea. It's the sort of discovery that takes place once in a career. "This is essentially by-catch," says Buffington, "Nontarget taxa were kept, but not just kept—they were actually mounted and labeled."

Unlike many other natural history specimens, insects are extremely durable if stored properly and well maintained. "Due to their cuticle, they preserve remarkably well," he says. "They can be 150 years old and you can still use them in primary research. As long as they've been kept dry and out of the sun, they're good to go."

These specimens were not just intact, they were pristine. When he left Logan, Buffington loaded them into the van, covered them carefully with a blanket, and continued driving. There were thousands of miles of road still ahead of him: northwest to Boise, Idaho, where he planned to pick up more samples; to Corvallis, Oregon, to collect more specimens still. A few days later, he added millions of Madagascan insects from the California Academy of Sciences collection. The broiling heat of Arizona and New Mexico still lay ahead, as did Texas, Mississippi, Tennessee, and then home.

In total, Buffington borrowed about five thousand wasps from the Bee Lab collection. Worldwide, there are an estimated three thousand species of cynipid wasps. He estimates there are hundreds of undescribed species in the drawers, which now await study in his lab in Washington, DC. "I'd probably say 250 new species, and probably two or three new genera," he says. That no one knew they were there in Logan is another example of how the taxonomic impediment slows progress.

"If life was beautiful and perfect, and every single specimen in the world was databased in one massive centralized computer system, I would've known they were there," says Buffington. "We just don't have the manpower to populate those databases. We're still very much in a nineteenth-century mode of discovering collections. You'd think since the collections are man-made these species would already be known. But they're not."

Instead, Buffington's work is part of a never-ending continuum. It depends on the work of others who came before him, and it will never be finished. "It's hard to imagine doing the kind of research I do—frankly it would be impossible—without the collections," he says. Even the best-tended collections contain a wealth of unidentified material. "There's something reassuring about this, that there's still unknowns out there."

Yours truly — Taken Sept. 8/39 (38?)
On cliffs above the Pastaza about 3 miles
to the north west of Baños — The two
young folks are Segundo Velastegin and
his 16 yr. old sister Rosario. The Velastegin family have been my helpers
for the past 16 years.   Mac.
              (Wm. Clarke – Macintyre.)
(From reverse of photo.)

12

The Biomimic:
The Lightning Cockroach
(*Lucihormetica luckae*)

Peter Vršanský is hunting for South American cockroaches. But he's not hunting in South America. Instead, he's in the Blattodea collection at the Department of Entomology in the National Museum of Natural History in Washington, DC.

From the box in front of him, which contains about a hundred specimens, Vršanský picks up a single cockroach. Turning it over in the dim light, he inspects its nut-brown body. It is, he says, a luminescent species of cockroach from Ecuador—otherwise known as a glowspot cockroach. In 2012 Vršanský, a researcher at the Slovak Academy of Sciences in Bratislava, named it *Lucihormetica luckae*, or the lightning cockroach.[1]

On top of its head is a dark brown ovate patch punctuated by two large pale circles, side by side. In fluorescent light they shine brightly, like two strange light bulbs. Vršanský calls them "great lanterns." The cockroach becomes a gleaming beacon. Beneath them, and only on one side, is another much smaller luminescent dot. On its back, two dark bands run along the outer edges of its body, and a third diagonal stripe runs along its back between them where the leathery, protective front wings are neatly folded, one on top of the other. When placed under fluorescent light, the bands remain dark, contrasting with the rest of the body, which glimmers.[2]

Vršanský had been in Washington, DC, for a monthlong visit to study a collection of fossilized cockroaches excavated from the Green River Formation in Colorado. He has previously described several cockroach species from fossilized prehistoric specimens, including some that came from the Green River site. On his last day in Washington, Vršanský visited the Department of Entomology to inspect the Blattodea collection—dead specimens of the still-living cockroaches. There-old but unnamed-he found the lightning cockroach. Its luminescence is a protective strategy, says Vršanský—an example of biomimicry. With its glowing lanterns, the cockroach has evolved to closely resemble the *Pyrophorus* beetles, which also have bioluminescent markings that gleam at night. Known as click beetles, they are toxic to predators. The cockroach has no natural defenses of its own. Instead it innovates, impersonating a click beetle. At dusk, unable to distinguish between the cockroach and a toxic *Pyrophorus* beetle, a hungry bird might opt to bypass the cockroach.

The single known specimen was collected on May 5, 1939, by William Clarke-Macintyre,[3] an enigmatic American collector who spent thirty years in Ecuador as a field collector. Clarke-Macintyre lived in Baños in Tungurahua Province, at high altitude, in the shadow of the Tungurahua volcano. For several decades, beginning in 1922 when he moved to Baños from New Jersey, he lived on the edge of the vine-choked jungles of central Ecuador, frequently disappearing into the overgrowth for months at a time accompanied by a dedicated team of native collectors. From the rainforest, he gathered specimens for American biologists. Printed at the top of his personalized stationery was the following: "William Clarke-Macintyre ~ Field Collector of Natural History Material ~ Collect In Any Order ~ By Special Arrangement." And in the top right corner of the page, in smaller type: "Now Collecting on Eastern Slopes of Andes, Ecuador, S.A."

A photograph from about this time shows Clarke-Macintyre bearded and smiling, seated midslope, wearing a flat cap and a jacket. He is fifty-eight. In the distance behind him another steep incline rises, and behind that a mist-capped peak. A long-handled collecting net rests in the crook of his arm. A girl sits beside him, her face indistinct. A little farther down the slope is a boy with his chin in the air. Written beneath the photo: "Yours truly—Taken Sept. 8/39 (38?). On cliffs above the Pastaza about 3 miles to the north west of Baños. The two young folks are Segundo Velastegin and his 16 yr. old sister Rosario. The Velastegin family have been my helpers for the past 16 years. Mac. (Wm. Clarke-Macintyre)."

From his home in Baños and across the Cordillera Oriental, Clarke-Macintyre collected more than just cockroaches. An online search of the Integrated Digitized Biocollection database shows more than 1,700 specimens in participating biorepositories around the world. The earliest was collected in 1917, the last in 1948. He covered an incredible amount of territory. His northernmost specimen, a bird called the chestnut-crowned antpitta (*Grallaria ruficapilla*), was collected from Quindío Province, Colombia, in April 1942; the southernmost, a velvet-purple coronet hummingbird (*Boissonneaua jardini*), was obtained in southern Peru, near Arequipa, in May 1937. For context, if one drove from Quindío to Arequipa today via the Pan-American Highway, the distance traveled would be 2,371 miles. It would take fifty-seven hours of nonstop driving on a road that didn't exist yet when Clarke-Macintyre was operating. And he collected everything: bats, bees, opossums, poison dart frogs, primates, rabbits, snakes, spiders, spiny-tailed rats, squirrels, toads, tree frogs, and weasels. He collected birds without interruption. Mostly, though, he specialized in collecting insects, especially Odonata, or dragonflies and damselflies, and Tipulidae, or crane flies.[4] He searched the headwaters of the Amazon and its tributaries for new species—along the Río Anzu and the Río Arajuno. He headed into the high country, beyond the moss-covered forests to the Río Blanco above Yungilla, and to Runtun on the northeast shoulder of Tungurahua. He canoed down the Río Pastaza to Abitagua to collect in the wet, swampy places there. He climbed above the snowline—like passing through a white wall into another world—to the treeless páramo of Tungurahua.

For decades he sent boxes filled with insects to Clarence Kennedy,[5] an entomologist at Ohio State University, charging five cents per specimen. Clarke-Macintyre and Kennedy have left behind decades of correspondence—a courtly friendship centered on insects. On September 1936, before embarking on a five-month expedition to the Río Jacun Yacu, Clarke-Macintyre wrote:

The territory where I expect to be working during Dec, Jan and Feb, I know has never been collected for entomological material, and should turn up a lot of very interesting material in all orders. I'm taking a good supply of vials, the alcohol, glycerine, perchloride of mercury and acetic acid, preservative sol., mailing tubes, etc, and will take special care in packing the larvae with bits of grass in the bottles and hope that they'll reach you in good condition. I've got to collect in a great many families on the trip—Tipulidae, Odonata, aquatic Hemiptera, ichneumons, asilidae, membracidae, coleoptera and lepidoptera, besides getting a collection of reptiles and amphibians for the Field Museum and fleas for Karl Jordan, so I'll be pretty busy but I surely have a splendid field force—6 very good native collectors and myself and my injun boy who now has been with me about 3 yrs so I hope for better results than ever before.

In 1939, about the time Clarke-Macintyre collected the lightning cockroach, Ecuador was politically unstable. Eventually the unrest affected his collecting efforts. In January 1939, while attempting to enter the Cordillera Oriental, Clarke-Macintyre was detained by the local military, who suspected him of being a Peruvian spy. He wrote to Kennedy:

When I reached Quito I went to the American Legation to complain of the treatment that I had received in the Oriente and the fact that the Army officers, to a man, seem to hate the guts of a collector, especially if he is an American. (Confidentially; if my name happened to be Giovanni d'Spaghetti, or Fritz von Swienhund and my Passport bore the seal of the Gov't of friend Hitler or Bro. Mussolini instead of that of the Land of the Free and the Home of the New Deal, I would have had no difficulties.)

Prevented from venturing farther afield, Clark-Macintyre stayed in Baños, hunkered down in the shadow of the volcano. At more than five thousand meters, Tungurahua is an imposing, broad-shouldered cone, looming over Baños like a silent green deity. In the Quichua language, Tungurahua means "Throat of Fire." On October 6, 1938, he had written to Kennedy:

My plans for the immediate future are rather vague. The political situation in Ecuador is in such a shakey state that I don't like to venture far from Baños until affairs settle down. It's almost impossible for a foreigner, right now, to obtain a passport to go into the Oriente. Brown and I are planning trips to Paramba, in N.W. Ecuador and to Loja, South Ecuador, but neither of us cares to start on such a trip until the coming Presidential election is over and calmed down. In the meantime, we will probably mark time collecting on the slope of Tungurahua and around Baños.

Weeks of turmoil slowly passed, and Clarke-Macintyre searched the Throat of Fire for specimens. On May 1, 1939, he collected several species of hummingbird from around Baños. Four days later, on the volcano that loomed above the town, he bent to retrieve the cockroach from the ground.

In 2010 Tungurahua abruptly exploded.[6] Depending on its range, the lightning cockroach might have been extinguished as a species in that single explosive moment. Baños, at the foot of the volcano, was quickly evacuated as lava poured down the slopes. The volcano has remained active since then, erupting periodically. The current status of the cockroach is unknown. Perhaps there is only one specimen left in the world—still on loan to Vršanský at the Slovak Academy of Sciences in Bratislava.

For more than twenty years, David Furth was the collections manager for the National Museum of Natural History's Department of Entomology. Recently retired, Furth is certain the collections still contain unnamed cockroach species—he estimates they hold ten thousand to twenty thousand cockroach specimens. But, says Furth, the cockroach collection is largely inaccessible now, a common impediment to taxonomists. "Part of our collections have been closed because of lack of staff, and lack of hiring more staff," he says. "Some call it deactivation. In our combined entomology department, and that includes our partners USDA and the Department of Defense, nobody has an interest in cockroaches, so all of the Orthoptera orders, including roaches, are kind of closed in a way."

In recent decades, Furth estimates that the Department of Entomology has lost almost half of its collections staff. It is impossible to borrow material from the deactivated parts of the collection. Although still well maintained, it has been placed in deep storage—warehoused indefinitely. Mothballed. If a scientific collection is like a library, deactivation is like taking the books from the shelves, packing them into boxes, and sealing the boxes: the library is effectively closed. In August 2015 Vršanský embarked on a month-long field expedition to Ecuador. Two separate teams of researchers disappeared into the same overgrown jungles that Clarke-Macintyre had scoured for samples and began searching for luminescent cockroaches in the wild. It rained every day. At higher altitudes they were constantly wet and cold. But the expedition was successful. "By now we have thirteen more new species just from Ecuador," says Vršanský. "Each of them is represented by a single specimen."

About one hundred miles northeast of Tungurahua another volcano, known as El Reventador—the Destroyer—rises from the humid jungle floor like a green pyramid. During the expedition, Vršanský searched for specimens on El Reventador and other nearby volcanoes. Every night for a month he walked upslope in the rain, sifting through damp leaf litter by flashlight, surveying the volcano. There on the slopes, in the dark, wet places, he saw the scuttling cockroaches, like little secret gleaming lanterns.

13

Sunk beneath the Surface in a Sea of Beetles: Darwin's Rove Beetle (*Darwinilus sedarisi*)

On August 24, 1832, HMS *Beagle* dropped anchor at Bahía Blanca, a deep natural harbor in present-day Argentina. On board was a twenty-three-year-old naturalist, Charles Darwin. He had been at sea since December 27, 1831, when the *Beagle* left Plymouth. Darwin had spent most of those months incapacitated with seasickness. During one bout of nausea, staring sadly down at a long, slow inescapable swell unfurling below him, he wrote, "This & three following days were ones of great & ceaseless suffering."

A few days before arriving at Bahía Blanca, Darwin had sent his first shipment of specimens home to Cambridge. Among them were four bottles of animals in preservative, rocks and tropical plants, several marine animals, and many, many beetles. On the coast, at Bahía Blanca, Darwin continued collecting specimens. Among the material was an unusually large species of rove beetle with a long, segmented body and an iridescent blue-green head. There, too, in the sandy ground near the harbor, he found numerous fossilized animal remains, which he described in detail in *The Voyage of Beagle*. He sent everything back to England.[1]

But the beetle disappeared. Then in 2012, 180 years after Darwin collected it in Argentina, Stelios Chatzimanolis found it again.

"It's an impressive beetle," says Chatzimanolis, an entomologist at the University of Tennessee at Chattanooga. First, for a rove beetle it's large. With its long, flexible segmented abdomen, it closely resembles a large and brightly colored earwig, which it is not. Its size sets it apart, but so does its appearance. Most rove beetles, he says, are dark brown or black. This one has a glittering blue-green head. Other key morphological characteristics stand out too. "It has what we call serrate antennae, like a saw," says Chatzimanolis, "which is not seen in other species or in other genera in the subgroup it belongs to."

Coleoptera, the animal order that includes all beetles, consists of more than 400,000 known species. By comparison, the entire Chordata phylum, which encompasses all animals with a backbone—fishes, amphibians, reptiles, birds, and mammals—numbers around only 65,000 known species. Within Coleoptera, rove beetles are members of a single family of beetles: the Staphylinidae. By itself it's an enormous group of beetles. There are about 60,000 named species of rove beetles; no one can possibly know them all. Taxonomically the Staphylinidae family is further divided into several subfamilies; the subfamilies are separated into tribes; the tribes are split into subtribes; and subtribes are then divided into genera and species.

Chatzimanolis is an insect systematist, and he is not a generalist. He's like a cartographer who, when asked to map out a sprawling megacity, instead traces in the minutest detail every building and street in a neighborhood on its outskirts and leaves the rest of the map totally blank. He specializes in a subtribe of rove beetles called Xanthopygina. In February 2012 Chatzimanolis was working on a revision of the *Trigonopselaphus* genus—a relatively small group of neotropical rove beetles contained within the Xanthopygina subtribe. When taxonomists embark on the revision of a genus, they often discover new species. They find discrepancies: specimens that were misidentified and placed into the wrong genus are reclassified; several named species are found to be a single species, which has been described again and again by different taxonomists. Species that once were named and then combined with others are resur-

rected. For the revision, Chatzimanolis requested all the unspeci-
fied *Trigonopselaphus* material from the Natural History Museum
in London. The specimens arrived by mail in a small box. Inside
the box were twenty-four pinned beetle specimens, all supposedly
belonging to the *Trigonopselaphus* genus. Among them, misidentified, was the specimen Darwin collected at Bahía Blanca in 1832.

Suddenly the beetle had resurfaced. It's difficult to overstate the
importance of finding an original Darwin specimen, collected during the *Beagle's* first voyage.

But where had it been for 180 years?

Darwin was an avid coleopterist. He began collecting beetles when
he was a child. In his autobiography he recounted hunting beetles
while he was still a student at Cambridge: "I will give a proof of my
zeal: one day, on tearing off some old bark, I saw two rare beetles
and seized one in each hand; then I saw a third and new kind, which
I could not bear to lose, so that I popped the one which I held in
my right hand into my mouth. Alas it ejected some intensely acrid
fluid, which burnt my tongue so that I was forced to spit the beetle
out, which was lost, as well as the third one."[2]

Perhaps he was reminded of that moment as he crouched at
Bahía Blanca to cup the sinuous, fast-moving green-headed rove
beetle in his partly closed fist. Decades later, in an 1858 letter to
botanist J. D. Hooker, Darwin wrote, "I feel like an old war-horse
at the sound of a trumpet when I read about the capture of rare
beetles."

Regardless, at some point the beetle Darwin collected in 1832
went missing. It disappeared amid an enormous collection of millions of other insects, built over centuries. By 1987 it was listed as
lost in "Darwin's Insects: Charles Darwin's Entomological Notes."[3]
Darwin had even assigned it a specimen number: 708. But it was
gone. Searching for it would be like looking for a specific stalk in a
towering haystack. "The Natural History Museum in London is not
a small place," Chatzimanolis says. "They have millions and millions
of specimens. If you misplace something, the chances of finding it
there again are near zero."

But in 2008 University of Copenhagen entomologist Alexey Solodovnikov did just that. He found the beetle again. A fellow rove beetle expert, Solodovnikov was looking through unsorted Staphylinidae material at the Natural History Museum in London when he saw it. After briefly inspecting the beetle, he moved it, tentatively placing the specimen incorrectly with other unsorted *Trigonopselaphus* material. It remained there for four more years, until Chatzimanolis requested specimens and it was sent to him by mistake.

"It came to me as unsorted *Trigonopselaphus*," he says. At that moment Chatzimanolis was one of just a handful of people in the world—perhaps the only person—who could tell at a glance that the beetle was misplaced. "I was able to easily figure out that this was not *Trigonopselaphus*," he says. "This is something new." Then he realized it was a Darwin specimen.

Finding the beetle and realizing it was a lost specimen collected by Darwin was just the first step in a much longer journey. He put aside the revision he was working on and concentrated instead on describing the lost beetle—its unusual serrate antennae curving away from its gleaming head.

"Being atypical by itself does not guarantee that a beetle is going to be described as a new species," he tells me. "The biggest hurdle we have to jump when we try to describe a new species is to get at where it belongs. That's the hardest job we have as taxonomists. We first have to learn everything that has been described so that when we see something new we'll be able to recognize it as new and be able to describe it."

In February 2014 Chatzimanolis gave the beetle the scientific name *Darwinilus sedarisi* and published his findings in the journal *ZooKeys*.[4] It is named for Darwin and for the writer David Sedaris, whose audiobooks Chatzimanolis listened to while writing the description. Before naming the new species, he searched extensively for another specimen—a paratype to describe alongside the holotype. Eventually he found one, part of the Museum für Naturkunde collection in Berlin. The second specimen is as mysterious as the first. It is partly dissected—damaged, really. The dark brown segments of its abdomen hang apart sadly like the bellows of a broken accordion. The tips of its damaged folded wings are visible. It is in poor condition but, unmistakably, it is another *Darwinilus sedarisi*. It was collected at Río Cuarto in Córdoba, Argentina—almost five hundred miles to the north of Bahía Blanca—by a collector named Breuer. The details are gone. Breuer must have collected it before 1935, when it was mentioned in a book by German entomologist Walther Horn. For all Chatzimanolis knows, it could be two hundred years old. If maintained properly, insect specimens can last for centuries. The entomology collection at the Natural History Museum in London includes pressed specimens of insects that date back to the 1600s. The oldest known pinned insect specimen—a Bath white butterfly (*Pontia daplidice*), resides at Oxford University Museum of Natural History as part of the Hope Entomological Collections. It was collected in 1702. "If you put a pin through them and let them dry very well so they won't get any fungi, they can survive like that for a long, long time," says Chatzimanolis.

Until recently, specimens were stored in naphthalene—a suspected carcinogen—to prevent damage by insect pests. Today entire entomology collections are periodically frozen at extremely low temperatures to kill any insects and their eggs. Even though insect specimens tolerate storage well, they require constant maintenance. "If you have a specimen in a museum in a box in a cabinet, you cannot assume that it will be well and survive for many, many years," he says. "It actually physically requires somebody to take care of that specimen."

In the 180 years since Darwin collected the rove beetle, the place it came from has changed drastically. The landscape has been transformed by humans. It has been shaped, cultivated, deforested—altered in profound ways. Chatzimanolis doesn't know if the beetle still exists. After struggling to find specimens in other collections, he suspects it is extinct. "The type locality has been converted to agricultural fields," he says. "If this is a species with a narrow distribution, it's gone."

Finding an undescribed species in a museum collection is not an unusual event, says Chatzimanolis. When, by some estimates, only one in four living insect species has been described and named, it's

an expected outcome. Millions of insect species still await description. An unknown number wait in natural history collections. "I'm
sitting in my office" he says quietly, "and behind me is a twenty-
drawer cabinet that probably contains more than a hundred new
species."

14

The Spoils of a Distant War:
The Congo Duskhawker Dragonfly
(*Gynacantha congolica*)

Completed in 1910, the Royal Museum for Central Africa sits solidly on a hilltop in Tervuren, Belgium, on the outskirts of Brussels, looking like an imposing neoclassical slab. The museum was built by King Leopold II, who died in 1909—too soon to see it opened. The museum had a singular purpose: to showcase the spoils of Leopold's bloody exploits in the Belgian Congo, the largest country in sub-Saharan Africa, now known as the Democratic Republic of the Congo.

Between 1885 and 1908, the Belgian king oversaw a murderous colonial campaign, exploiting the Congo's natural resources to finance the construction of palaces and civic projects at home.[1] First, ivory was harvested from elephants. Tons of it was sent to Europe, and elephant carcasses littered the African landscape like shipwrecked hulls. Later Leopold built a thriving rubber industry that was powered by Congolese slaves. An estimated ten million people died in the process. If Belgian supervisors decided slaves weren't working fast enough, they would routinely chop their hands off with an ax.

The vast colonial collection—bloodstained as it is—remains archived mostly at the Royal Museum for Central Africa, and elsewhere too. And it still contains surprises.

In December 2015 Dutch entomologist Klaas-Douwe Dijkstra published a monograph on African dragonflies in the journal *Odonatologica*.[2] Odonata is the order of insects that comprises dragonflies and damselflies. In this publication Dijkstra and his coauthors named sixty new species at once—the greatest number of dragonflies and damselflies named at one time since 1886. "For African dragonflies, it adds one to every twelve that were already known," says Dijkstra, who completed the work at Stellenbosch University, near Cape Town in South Africa. With a single publication—it is a compendious 236 pages—the ranks of known African Odonata have increased by almost 10 percent.

Many of the new species were discovered through fieldwork in countries across western and central Africa, including Angola, Cameroon, Gabon, Uganda, the Democratic Republic of the Congo, and others. It has taken Dijkstra and his coauthors Jens Kipping and Nicolas Mézière several years to complete, traveling across the midsection of the African continent. But some of the new African species were discovered far from Africa. Instead they were in Tervuren, twenty miles east of Brussels, among the colonial artifacts at the Royal Museum for Central Africa. A century earlier, as unrest spread across the Congo basin, field collectors were busy collecting the unknown dragonfly biodiversity and sending specimens back to Belgium along with everything else. There, in their unknown thousands, they were placed in drawers and left.

One of the new dragonfly species is *Gynacantha congolica*, the Congo duskhawker. It's a delicate three-inch-long dragonfly with a wide lime green head and finely venated yellow wings. Dijkstra caught one in the Democratic Republic of the Congo during his fieldwork. Older specimens exist. In March 1899 a field collector named Waelbroeck netted two female specimens near Kinshasa, the capital city. During Belgian colonial rule, Kinshasa was known as Leopoldville. Those specimens are now almost 120 years old, and they're part of the entomology collection at the Royal Belgian Institute of Natural Sciences in Brussels, which is one of the largest insect collections in the world, with approximately seventeen million specimens.

Dijkstra captured the holotype of *G. congolica*—a male specimen—during fieldwork in the northern half of the Democratic Republic of the Congo in August 2010. He was standing in a flooded forest near Isangi, a small rural village on the banks of the Congo River, which breaks apart there like a frayed brown thread—first into two forks that continue north, then into a profusion of narrow braided channels snaking and elbowing north through the forest, briefly rejoining, then unraveling again. Brown water moves slowly through a green, flooded forest; tree trunks grow at strange angles in the water.

In the monograph there is a photo of Dijkstra working at Isangi. On the far side of the river he sweeps his capacious net through the air, back bent against a thick green wall of foliage. The net is filled with air. Four boys watch Dijkstra from their canoe, idling in the brown water of the river channel. Dijkstra belongs, he says, in the field—in difficult places. "This is a human being's natural state," he says. "We're hunter-gatherers much more than we're farmers or industrialists or anything else. This is how we get our mental sustenance."

Worldwide, there are about seven thousand Odonata species. Most common in tropical habitats, they are found in every faunal region on Earth except Antarctica.[3] With their striking colors and large body size, dragonflies have always attracted entomologists. In a 1700 letter to English naturalist Martin Lister, the French entomologist François Poupart wrote about the dragonfly, "It is wonderful how it rends and cuts the air, making a thousand whirlings with its extraordinary quickness."

In Africa alone there are almost eight hundred named species. Often they are sentinel species—they can be used to determine the relative health of the ecosystem where they're found. But only if we know them. Dijkstra estimates he now has captured more than 80 percent of African species in the wild, patiently sweeping his net back and forth in wet, swampy places like Isangi. It's elemental: a long, unbroken sweep of a net through air. No technology or advances can replace it. Since an initial field expedition to Ghana in 2000, he has returned to Africa every year, collecting specimens from across the continent. In total, he says, he has spent more than a thousand days in the field, in twenty countries.

Most surprising of all, many of the novel species Dijkstra named are conspicuously different. A photograph is enough for any expert in African dragonflies to assess them and know they are new to science. They are distinct and brightly colored—banded and striated with pink and purple, their cobalt blue abdomens streaked with tangerine—like whirling fragments of neon against the turbid brown water of the Congo. As Oliver Goldsmith wrote in 1791 in *An History of the Earth, and Animated Nature,*

Of all the flies which adorn or diversify the face of Nature, these are the most various and the most beautiful; they are of all colours, green, blue, crimson, scarlet, white; some unite a variety of the most vivid tints, and exhibit in one animal more different shades than are to be found in the rainbow. They are called, in different parts of the kingdom, by different names; but none can be at a loss to know them, as they are distinguished from all other flies, by the length of their bodies, by the largeness of their eyes, and the beautiful transparency of their wings, which are four in number. They are seen in summer flying with great rapidity near every hedge, and by every running brook; they sometimes settle on the leaves of plants, and sometimes keep for hours together on the wing.

Specimens of many of the new species had waited at the Royal Museum for Central Africa for almost a century. "The collection is completely packed with masks and canoes and everything-animals, dragonflies, artifacts of any sort, basically anything," says Dijkstra. "If you look at the dragonflies—but you could do the same for any other animal or plant group—basically what you have is just drawers stuffed full of stuff."

Much of it has not been cataloged at all. No one knows what it is. "Often it's not even been touched by a scientist ever," says Dijkstra. At some point Frederic Charles Fraser, a prominent English dragonfly systematist, had worked his way through parts of the collection, labeling and identifying some of the specimens. But Fraser died in 1963. "Maybe if you're lucky an identification label by him has been put in with these specimens, But in many cases not. You have these big drawers. The moment you take off the lid the material just pops out, because it's been forced in there—actually pressed down to keep the stuff in. The moment you take off the lid, it overflows."

The specimens aren't pinned in orderly rows in their drawers as they are at other institutions. Instead, each dragonfly is sealed inside an individual packet. There are packets everywhere. There is no order—no labels. Everything is in complete disarray. "You pop open a drawer, and there are hundreds if not thousands of specimens in there," Dijkstra says. "You're just dealing out these piles: that's this species; that's that species; that's that species. And then you're left with these oddballs."

The oddballs are many: *Gynacantha congolica*, first collected in 1899, was left in a packet for 116 years. Dijkstra named *Chlorocypha granata*, the garnet jewel, which had first been collected in 1935. He added *Anax gladiator*, the swordbearer emperor, collected in 1951 by Charles Henri Seyer near the Congolese border with Zambia; *Notogomphus intermedius*, the Katanga longleg, collected a year earlier by Seyer; and *Neodythemis katanga*, the Katanga junglewatcher, netted by G. F. Overlaet in 1933, in Kapanga. Dijkstra has named them all. He gave one species the Latin name *Umma gumma*—the title of a 1969 album by Pink Floyd. Its common name is robust sparklewing.

Finally, they are all known.

Dijkstra's work has practical consequences. The International Union for Conservation of Nature (IUCN) maintains a list of endangered species—known officially as the Red List of Threatened Species. Each year the list is updated to include species that have become endangered and need protection. But a species can't be placed on the list if no one knows it exists in the first place. "Everything starts with that name. It's almost an act of registration," he says.

Many of the new species face habitat destruction. Their type localities—often wet, swampy places—have suffered from years of drought and climate change. Without names, vulnerable dragonfly species remain unprotected, and some just disappear.

I have spent considerable time looking at the photographs of the new species in Dijkstra's monograph. To my untrained eye they all look similar. But to a systematist like Dijkstra—an odonatologist—the differences between them are profound and conspicuous. The morphological differences somehow engage and activate parts of Dijkstra's brain that clearly don't work in mine, allowing him to determine their uniqueness without difficulty. "It's something taxonomists are good at, seeing a pattern, even if they can't immediately say why," says Dijkstra—just as Blackburn did with the overlooked squeaker frog.

Dijkstra has a friend who can differentiate between almost all the species of butterflies in the *Bicyclus* genus—a genus of almost one hundred butterflies found in the Afrotropical ecozone. "All of them are brown and very similar." Nonetheless, by means even he doesn't fully understand, his friend Oskar can tell them all apart. He can differentiate them—and compute the differences. He is the only person in the world able to discriminate them all. Some of Dijkstra's other friends have written a self-learning computer program to identify *Bicyclus* species. Over time, the program learns and becomes better at differentiating species. They named the program Oskar 2.

"After a while it could identify 97 percent correctly, even if they didn't actually know how it analyzed the differences," says Dijkstra. "It's pattern recognition," he says. "You pull open the drawer and you see it instantaneously. Sometimes you're pulling open a drawer that a dragonfly expert has pulled open before you, probably hundreds of times. When they opened that drawer, it just didn't click. But you open it and you see, 'Hey, there are labels saying there are just two species in this drawer. But I can see three.'"

To someone with those powers of discrimination, an unsorted drawer of dragonfly specimens collected almost a century ago from the banks of the Congo—or the outskirts of Kinshasa, or along the swampy Congolese border with Zambia—vibrates with meaning. With every new species he discovers, biologists understand more about the biodiversity of an underexplored ecosystem. Dijkstra's eyes dart quickly over the unidentified packets in the drawers, taking them all in. In some of the packets specimens are broken:

delicate old wings are crushed; legs have desiccated to dust in the drawer as a century passed by on the outside. But Dijkstra is looking, searching the visible contents of each sealed packet for visual clues—sorting, processing, endlessly computing.

"It's something that either you can do or you cannot," he says. "Some people are musical. Other people can do this."

He opens another drawer and spreads his hands across the front to stop the contents from spilling out onto the floor. Picking up another specimen, he holds it up to the Belgian light like a jeweler appraising a diamond, and networks of neurons in his brain begin to fire.

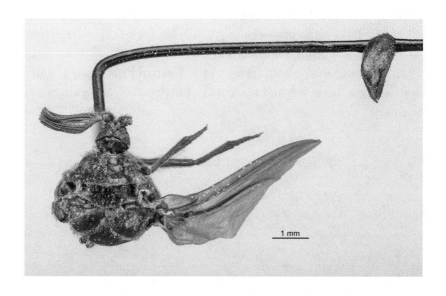

15

A Specimen in Two Halves: Muir's Wedge-Shaped Beetle (*Rhipidocyrtus muiri*)

In the summer of 1907 Frederick Muir, an entomologist at the Experiment Station of the Hawaiian Sugar Planters' Association in Honolulu, was on the island of Borneo in Indonesia. He stood in the wet heat of West Kalimantan, enclosed on all sides by thick mangroves. Picking his way through a wall of screw pines eight feet thick, he stepped cautiously into a bamboo-thatched canoe, pointed its nose upstream in muddy water, and began to paddle eastward, into the interior. "The whole aspect is very tropical, especially where the rattans shoot their graceful tops above the trees," wrote Muir.[1]

He was in Borneo to look for a parasite—a weevil parasite he had never encountered before. For all he knew, he might have seen it already and missed it. He could have walked straight past it a week earlier. After all, how would he know? By then Muir, an Englishman, had already been searching for a year—in China, Macao, and Singapore, through jungles everywhere. Back in Hawaii, sugarcane growers were fighting what seemed like an unwinnable war. Their foe was the sugarcane borer (*Rhabdoscelus obscurus*), a voracious species of weevil: a small walnut-colored beetle with a long, down-curving rostrum like a slender black beak. It had been introduced to the islands accidentally. The weevil lays its eggs in the sugarcane, and its larvae eat unstoppably through the cane, riddling it with holes.

Every year 10 to 15 percent of the sugarcane crop was being lost to the weevil. If Muir could find a parasite to keep its larvae in check, he might save the crop. This technique of biological control— introducing a naturally occurring parasite to another ecosystem so that it can control pest species—is employed all the time today. It's a technique Matthew Buffington uses. One of the reasons Buffington is interested in cynipid wasps is their potential to control agricultural pests that decimate crops. But in 1906, when Muir left Honolulu to find a parasite for the sugarcane borer, he was a pioneer.

Muir didn't find the parasite he was looking for in Borneo, but he collected many other insect specimens while he searched for it. One of them, obtained from among the mangroves and screw pines of western Borneo, was a small wedge-shaped beetle that he collected in August 1907. There is only one specimen—a male, now part of the entomology collection at the National Museum of Natural History. The underside of its label reads "Found by F. Muir on flowers." Zachary Falin, entomology collections manager at the University of Kansas, named the beetle in 2014. He called it *Rhipidocyrtus muiri.*[2] It had waited 107 years to be named.

In the intervening years the beetle had been disarticulated. Most of its front half was intact, but the rest had been dissected and mounted on a series of glass microscope slides. This isn't a barrier to identification, but at some point the composite parts had become separated. By itself each half was worthless. They spent decades separated. The front portion of the beetle is mounted on a pin. To a noncoleopterist like me, it looks more like a strange brown nut. It's dusty, as any old object would be. I can't really tell one end from the other. The solitary landmark is a single elegant twisted antenna, or a flabellate lamellate antenna—*flabellate*, meaning "fan-shaped," and *lamellate*, meaning "arranged in flattened plates." Falin is a world expert in the Ripidiini—the tribe of beetles the beetle belongs to. In turn, the Ripidiini tribe belongs to a larger family of about 450 species, the Ripiphoridae, known as wedge-shaped beetles. They are very small squat-bodied, high-shouldered insects. Their elytra — or wing cases — are tiny too, leaving their rear wings exposed, which makes them look even less like beetles. Within its larger family, the Ripidiini tribe is less well known. There are only sixteen genera, most housing just a species or two. "These things are rare," Falin says. "They're morphologically just bizarre."

As a graduate student in the mid-1990s, realizing the group was in disarray and poorly characterized, Falin began to borrow Ripidiini specimens from other museums. "I contacted fifty or sixty museums around the world," he says. "I received thousands of these things in the mail in the first few years of my graduate career."

Their life cycle is profoundly unusual too. "As far as we know from only one or two cases, they are known to be internal parasitoids of cockroach nymphs," says Falin. "Young cockroaches harbor these as internal parasitoids, and after a certain period the beetles on the inside pop out, sort of like *Alien*, killing the roach, and then pupate, turning into adults."

In other words, for most of their life cycle they're hidden. They're tunneling through the darkness in the parasitized body of a cockroach, invisible in the leaf litter on the jungle floor. When the beetle finally emerges and pupates into its adult form, it's still strange-looking. "The females are larvae form, which means they have no wings. They look like larvae. They're very degenerate," Falin says. "The males have virtually no mouth parts. They don't eat. They have gigantic eyes and gigantic antennae. It's pretty obvious that the only thing they're good for is to find a female." Even its elaborately fanned antennae, he says, are designed for that purpose, covered as they are with sensors to detect the pheromones of potential mates wafting through the humid air. They hatch. They breed. They die. All in a matter of days.

When Muir arrived in Borneo in 1907, he was entering the green heart of darkness. It was a wild and trackless place. For decades Europeans had been captivated by sensationalist and mostly inaccurate accounts of cannibalism like Carl Bock's *Headhunters of Borneo* (1881). An enormous island—the third largest in the world—Borneo is divided between three countries: the largest part, encompassing the southern half of the island, belongs to Indonesia; Malaysia controls the northern part of the island; on the northwest coastline the tiny independent state of Brunei occupies an area smaller than Delaware. Borneo is home to some of the world's oldest rainforests and is recognized by Conservation International as one of thirty-five biodiversity hot spots worldwide. Many species are found there and nowhere else.

Among such bewildering biodiversity, *Rhipidocyrtus muiri*—a small nut-brown beetle with long, twisted fanlike antennae—went unnoticed. Back in 1996, for his studies Falin needed to see as many ripidiine beetles as possible. The search took him to the entomology collection at the National Museum of Natural History at the Smithsonian Institution. "They had plenty, but they also had some specimens that were missing," he says. "I thought, whatever is missing is probably the interesting stuff."

Several decades earlier, another researcher had borrowed a large part of the collection and had not yet returned it. Specimens go missing from museum collections all the time. Sometimes even holotypes—marked with red tape or ribbon to denote their importance—disappear. But Falin was lucky. Within a few months the missing ripidiine specimens were in Kansas.

"Among that material was this kind of torn up adult—one single specimen from Borneo," he says. "It was very striking, but it was incomplete. However cool it was, I just couldn't do anything with it. I couldn't use it for my analysis, and I couldn't describe it. Half of it was gone. So it just sat, and sat, and sat, and sat."

In 1888 a piece of infested sugarcane arrived at the Department of Agriculture in Washington, DC, from Hawaii—then still known as the Sandwich Islands. It had been sent at the request of Kalakaua, the Hawaiian king. "The cane received at the Department proved to be infested by the larvae of a large Snout-beetle of the genus *Sphenophorus*, several species of which are known to bore into the stalks and roots of corn in this country," wrote entomologist Charles Riley in an 1888 report in *Insect Life*.[3]

Normally stout and woody, the cane was "completely riddled with the galleries of the larvae," Riley wrote. Adult weevils had laid their eggs in the cane and, on hatching, the grublike larvae had tunneled through it, leaving behind wide chambers filled with macerated cane fiber. The weevil was first described in 1835 by French lepidopterist Jean Baptiste Boisduval, working in Paris from specimens collected in 1827 by explorer Jules Dumont d'Urville during the first voyage of the *Astrolabe*. Dumont d'Urville had inadvertently brought infested cane back from New Ireland in Papua New Guinea. By then the weevil was in Tahiti too. Research suggests it was introduced to Hawaii in 1854 aboard the *George Washington*, a whaling ship, where it infested cane first at Lahaina and then throughout the islands. This was what had sent Muir to Borneo and elsewhere. For the Hawaiian sugarcane growers, the weevil was an ecological disaster. In many ecosystems, pests like the sugarcane borer are present but are kept in check by their natural parasites. When the borer was introduced to Hawaii it had no natural parasites, and it spread unchecked. Sugarcane growers reported picking weevils from the cane by hand. In one year, from just one plantation, workers picked twenty-seven thousand ounces of parasites from the cane—almost a ton of the tiny beetles.[4]

So Muir was sent across the world to find something that would take the weevil down. In Indonesia he was prevented from inspecting the sugarcane by an ongoing war between the natives and Dutch colonialists. Later, in the Maluku islands, he was almost blinded when a cobra sprayed its venom in his eyes. Traveling through Guangdong Province in southern China, a region hostile to outsiders, he pretended he was a doctor searching the hillsides for medicinal herbs. In Port Moresby, New Guinea, he contracted typhoid fever and convalesced for weeks in an Australian hospital bed. Nothing stopped him for long.

Though he couldn't have known it at the time, Muir was searching for *Lixophaga sphenophori*. Finally he found it in January 1908 on the island of Ambon. An unassuming little black tachinid fly,

L. sphenophori parasitizes an Indonesian species of beetle borer that lives on sago palms. He found it in New Guinea too, where it seemed to control the sugarcane weevil numbers. It might control the weevils in Hawaii too, Muir reasoned, if he could get it there.

In the end he was right. There were several failed attempts to transport the flies to Hawaii. Every time, dead flies arrived in Honolulu. On one occasion Muir tried taking a box of flies with him to Hong Kong, but they all died the day before he arrived. Eventually he had to set up relay stations, ferrying flies from New Guinea to Australia, breeding them, then taking their pupae to Fiji, where he bred them again. One leg at a time, he transported the flies across the world. Finally, in August 1910 Muir walked to the edge of the Hawaiian cane fields with a box in his hands, assessed the direction of the breeze, turned his back to it, and released the flies. The introduction of the fly to Hawaii worked. After four years on the project, Muir took a yearlong vacation before embarking on another search—to find a natural enemy of the cane root grub (*Anomala orientalis*).[5]

The pinned and disconnected front portion of *Rhipidocyrtus muiri* is tiny, a brown speck on a pin. When Falin tells me it's large, which he does several times, he means relative to other ripidiine beetles. In those terms it's gargantuan. "Whatever cockroach this thing is living in has got to be a big roach," he says. "It was striking in that regard."

But it's still only three or four millimeters long. It's easy to overlook. Despite this, Muir knew the beetle was unique when he collected it from a flower in Borneo. In 1918, eleven years later, he visited the National Museum and asked entomologist Eugene Schwarz to find the beetle for him. Ten years after that, during what was to be his final visit to the collection, Muir again asked curator Herbert Barber to find the beetle for him. When Muir died in 1931 the beetle was still unnamed. It remained on its pin—overlooked and unknown. During a 2011 research visit, Falin was in the back of the collections, seated at a microscope surveying ripidiinid beetle specimens. A few feet away, inspecting her own material, was University of Oxford entomologist Jeyaraney Kathirithamby. She is an expert in the Strepsiptera, a group of parasitoid insects that are often mistaken for Ripidiini beetles. Kathirithamby was holding some slides that had been stored incorrectly among the Strepsipteran holdings, like a library book returned to the wrong shelf.

"These aren't strepsipterids," she said, turning to Falin. "I think they're ripidiinids instead." "I took the slides from her," says Falin. "Instantly I realized, these are the missing bits from that giant bug I have at home. Somehow, sometime after the 1930s or 1940s, these dissected bits were misidentified and curated with a completely different order of insects. There's no way for anyone to have realized that unless Kathirithamby and I were sitting in the same room at the same time."

Together the three slides and the pinned front half of the beetle constituted a whole specimen. Finally Falin could describe it and give it a name. It had waited for more than a century. The hidden biodiversity in natural history collections is a reflection of how little we know about insect diversity in general. This is particularly true in biodiversity hot spots like Borneo where so much remains to be discovered.

Falin wonders if the beetle is still present on Borneo. He has never seen another example of it, despite examining thousands of specimens. To an expert in wedge-shaped beetles, it's a distinctive species. "Anyone who saw one, who knew anything about these things, would snap it up in a second," he says. "It just goes to show: there are things flying around in the jungle that we still have no idea about."

16

Mary Kingsley's Longhorn Beetle
(*Pseudictator kingsleyae*)

Coleopterist Max Barclay is seated at his gently cluttered desk in the Natural History Museum in London, holding a black beetle. It's as long as Barclay's thumb. It has implausibly long segmented antennae—they sweep backward and are longer than the beetle's tapered body, extending beyond its elytra. These long, curving antennae are the morphological feature that earns the beetle its classification as a longhorn. As collections manager at the museum, Barclay oversees a vast archive that includes about ten million beetle specimens. But the beetle in his hand is the only one like it there— a single specimen. It came from present-day Ghana in West Africa, and it has spent most of the past century stored in a box. In 2015 the beetle was named *Pseudictator kingsleyae* after Mary Kingsley, the woman who collected it in 1896.

Periodically during the past century, entomologists have removed the specimen from its box and inspected it. They have examined it meticulously, measured it, and lined it up alongside other examples of African longhorn beetles. No one could correctly identify it, although they've tried. "Various people have looked at it and attached labels to it, commenting that they're not sure what it is," Barclay says.

Worldwide, he says, there are approximately 25,000 to 30,000 species of longhorn beetles. Most are much smaller than the specimen Kingsley collected. Morphologically, it is different from related species in other ways too. "It's distinguished from other beetles by the thorax—the pronotum behind the head and before the abdomen—being very much swollen and distended, at least much more so than in the closely related genera," says Barclay. It most closely resembles the dictator beetles, a small genus that consists of thirteen distinct species. "It's called *Pseudictator* because it's most similar to that genus," he says. "It's a false dictator." The false dictator—the only known member of its genus.

Born in England in 1862, Mary Kingsley was a fearless explorer. She received no formal education—at the time, in Victorian England, educating women was considered unseemly. In particular, the sciences were not something a genteel young lady should busy herself with. Studying them might make her ill. Instead, when her younger brother Charles left to study law at Cambridge University, Mary remained at home to care for her ailing parents.

In 1892 Kingsley's parents died within a few months of one another. Suddenly she was freed from her obligations. Funded by a modest inheritance and self-educated from the books in her father's extensive library, she decided to travel alone to West Africa. She knew no one there; she had never even left England before. "To look at her one might think a puff of wind would blow her away," wrote her friend the writer Ethel Tweedie.

When Kingsley asked doctors for advice, they cheerfully told her that West Africa was the deadliest place on Earth—disease, violent natives, the numerous untold dangers of the bush. The journey there took two weeks by steamboat. The steamboat agents in Liverpool didn't offer return tickets: no one came back; eventually everyone died. Kingsley went anyway. Finally, in August 1893 the thirty-year-old stepped off the boat into the heat and the sprawling clamor of Freetown, in Sierra Leone. Kingsley made several expeditions, which she described in her popular memoir *Travels to West Africa*, published in 1897.[1] She was the first white woman to ascend

Mount Cameroon, which, at 13,250 feet, is the highest point in sub-Saharan West Africa.

Describing the Gold Coast—present-day Ghana—she wrote,

It is this way: you are coming home from a long and dangerous beetle-hunt in the forest; you have battled with mighty beetles the size of pie dishes, they have flown at your head, got into your hair and then nipped you smartly. You have also been considerably stung and bitten by flies, ants &c., and are most likely sopping wet with rain, or with the wading of streams, and you are tired and your feet go low along the ground, and it is getting, or has got, dark with that ever-deluding tropical rapidity.

Mostly Kingsley was interested in fish specimens and ethnographic artifacts. Even with insects invading her voluminous petticoats, she stopped at nothing to obtain them—she forded rivers, ascended mountains, and hacked her way through unexplored rainforests. During her years in Africa, Kingsley collected numerous specimens—fish, reptiles, insects, shells, and grasses. While in Ghana in 1896 she collected some beetle specimens, but only a dozen. Modern-day field expeditions return with vials filled with thousands of specimens; Kingsley collected twelve. *Pseudictator kingsleyae* was one of them.

In October 2014, as he does twice a year, Barclay went to the Prague Insect Fair, an event attended by thousands of entomologists from around the world. This time he took along the unidentified longhorn beetle. "I knew there was a French guy there who was the world expert on this particular group of African longhorn beetles," he says. The French guy was Pierre Juhel. "I stuck it in front of him, and he immediately said it was a new genus," says Barclay. "I suggested he should name it after Mary Kingsley, which he did."

P. kingsleyae belongs to the Cerambycidae subfamily of beetles, a cosmopolitan group of about thirty thousand species. Describing any new species is challenging enough: it becomes even more difficult when the unidentified specimen belongs to that large a group of beetles.

"Identification is a process of elimination," Barclay says. "When you've got a group with 30,000 species in it, you've got to apply a process of elimination that's going to eliminate 29,999 of them. You have to start eliminating from the known fauna: you know it's not a mammal; you know it's not a bird. That's always very obvious. You know it's not a butterfly, or a fly, or a wasp. So you've got it down to a beetle."

Even then, says Barclay, the hardest work still lies ahead. When approximately one-fifth of all known species are beetles, knowing a specimen is a beetle is not particularly helpful. "You've still got 400,000 possibilities. You've got 180 families of beetles. If you recognize the families you've got a chance of knocking it down to 20,000 or 30,000 possibilities; then you're going down through the subfamilies and the tribes and the genera. Each time you make an identification to a group—if you make the identification correctly— you've knocked out almost an order of magnitude at the beginning. Eventually you get it down to a manageable number."

Finally, the taxonomist is left with two possibilities: either a specimen is correctly identified to the species level or it remains unknown and must be formally described. Species discovery happens all the time in collections, especially from the drawers and cabinets that Barclay oversees. The beetle collection at the Natural History Museum in London is one of the most comprehensive of its kind. Every year it grows larger and larger—it is forever expanding. The collection is stored in more than 22,000 drawers. As it expands, it steadily becomes more powerful. Assembled over centuries, it contains more than 200,000 type specimens—the single specimens that are used as reference points to describe an entire species. Every year, says Barclay, more than a thousand novel beetle species are described from material in the collection. In September 2014, while digitizing unidentified material, Barclay found an old tattered box that contained twenty beetle specimens that explorer David Livingstone collected in the 1860s during an expedition to the Zambezi River.[2]

As collections manager at the museum since 2001, Barclay has described twenty or so new species himself. About a hundred new species have been named after him. He has assisted in naming hundreds more—by finding an unnamed specimen, recognizing it as new, then passing it on to a specialist in that family or tribe or sub-tribe who can describe it, as he did with Juhel and the longhorn beetle.

"I've got a box of beetles on my desk today," he tells me. "They're also longhorn beetles. There are three species in here that are new to the collection. One of them is new to science." The box sits amid papers and other specimen boxes strewn across Barclay's desk. He reaches to pick the beetle from the box, bending closer to inspect its label. "It looks like it was collected in 1912," he says slowly, strug-gling to read the words. "It's from Sierra Leone. It's got a label on it from a past curator in the 1950s that says 'species close to, but not' and then the name of another species. Somebody's noticed that it's different, but they haven't quite been able to follow it up. That happens all the time."

Even though the collection is comprehensive, it has its limita-tions. "We've got about 60 percent of known beetles here," Barclay says. "So, if I take a beetle down to the collection, I've got a rea-sonable chance of matching it. But if I can't match it, I don't know whether it's in the other 40 percent—which is in the Smithsonian, or the Paris Museum, or Copenhagen, or somewhere else—or whether it's a new species."

This is why the collections are so important. None of them is complete. But when a specimen becomes part of a collection it be-gins another life—a second life as a representative. Like many tax-onomists and curators, Barclay is frustrated by the notion that collections are antiquated and anachronistic. "People think of mu-seums, and they think of the last scene of *Raiders of the Lost Ark*," he says. "Someone has been fighting and struggling to get this trea-sure, and then they put it in a crate and stick it in a basement where there are rats and dust. The door shuts, and that's it. They think, 'Yeah, okay, now it's in a museum, so, you know, it's not going to go anywhere or do anything anymore.' I think it's slightly unhelpful that we have those views."

In June 1900, a few years after collecting the beetle in Ghana, Mary Kingsley died. She was thirty-seven. She had embarked on another expedition—this time to South Africa to collect fish specimens from the Orange River. Then in October 1899 the Boer War began. With widespread fighting between the British and troops from the Orange Free State and the Transvaal Republic, Kingsley volunteered as a nurse in Cape Town, tending to the injured Boer prisoners of war. "All this work here, the stench, the washing, the enemas, the bedpans, the blood, is my world," she wrote in a letter to a friend. Working in the wards of Simonstown Hospital, surrounded by the dead and the dying, Kingsley contracted typhoid fever.[3] She was buried at sea. Most of the specimens she collected remain at the Natural History Museum—fish collected from Cameroon and Gabon, and the beetles. "She was the first woman to get to the top of Mount Cameroon, the tallest mountain in West Africa," says Barclay. "She wanted to push the boundaries—not only of what was socially acceptable, but of what had been done."

The opportunities to collect a specimen like the one he's holding were few, Barclay says. Burrowed into the wood, it avoids light. Its adult life is brief—a few weeks at the most. "It's collected from an obscure part of West Africa where there really haven't been a lot of expeditions," he says. The single specimen collected in 1896 is the only known example of an entire species. "There might have been one opportunity in the history of humanity to have collected that beetle," he says. "It was given to Mary Kingsley, and she took it."

The Giant Flies

(*Gauromydas papavero* and *Gauromydas mateus*)

Julia Calhau pulled out a wooden drawer filled with flies. Giant flies. In 2009, while completing her graduate studies at the University of São Paulo in Brazil, Calhau was gradually becoming an expert in the Mydinae—a subfamily of flies that includes several genera of superlatively large flies.

Although she wasn't particularly interested in taxonomy, the number of species in the group was small enough that eventually she became proficient at telling them all apart. Calhau is a dipterist: she studies flies. When she began the *Gauromydas* genus was made up of just four species of large neotropical flies, one of which— *Gauromydas heros*, an enormous, oil-black monster—is the largest species of fly in the world. Fully grown specimens have wingspans measuring almost four inches. It is the king of flies; a thousand houseflies rolled into one.

But they are rare—in life, and in scientific collections too. The natural history of the *Gauromydas* genus is still mostly unknown. Buzzing above the Brazilian savanna, they might even be mistaken for small birds. They are not collected often: even the largest ento-mological collections might have just a handful of specimens—or perhaps none.

The male adult *Gauromydas heros* feeds on flowers, but the female adult doesn't seem to feed at all. Dipterists like Calhau suspect the flies live near ants — *Gauromydas* larvae have been found in the waste chambers of ant nests, feeding on the other immature insects they find there. Eventually, fattened on a diet of insects, a mature *Gauromydas* larva leaves the strange, dark ant-generated heat of the chamber, like an underground womb. Migrating upward, it digs through the soil, stopping about six inches from the surface to construct a pupation chamber for itself. Later the imago — the adult form — emerges from the ground: an enormous darkling thing.

In 1963, when he was twenty-one and recently graduated as a biologist, Nelson Papavero began working on the insect collection at the Museum of Zoology of the University of São Paulo. A native of São Paulo, Papavero eventually became a world-renowned expert on flies, organizing an extensive catalog of Diptera that is still used today. He was one of the scientists who erected the *Gauromydas* genus.[1] With a resident expert like Papavero, Calhau was surprised to find a drawer in the museum collection that was filled with unidentified material.

In the drawer, pinned alongside twenty or so other specimens, Calhau noticed one large dark fly in particular — long-bodied and solid, with intricately veined wings. "The specimen was beautiful and so nicely mounted, and it was part of the collection by Dirings," she says. Richard von Diringshofen was an amateur German-Brazilian entomologist: the museum obtained his personal collection when he died in 1987. To begin with, Calhau identified the fly as *Gauromydas mystaceus*, another species in the genus. Collected in Oriximiná in 1960, it had long, pointed tibial spurs on its hind legs. Entomologists believe that tibial spurs — sharp, curved protrusions that look like thorns — are a morphological adaptation that aid in digging and that insects work the spurs through the soil like picks.[2]

The fly's wings are a muted orange, engraved with a fine tracery of veins like mapwork. Its powerful black hind legs resemble a grasshopper's legs. A photograph shows a female, its body wrapped in a pair of red-brown wings the color of polished wood. In 2012 Calhau compared the fly with other museum specimens, which came to her on loan from other institutions in Brazil: from the Museu Paraense Emilio Goeldi in Pará State, and from the Instituto Oswaldo Cruz in São Luis in northeastern Brazil.

With the borrowed insects lined up in front of her, Calhau peered through a dissecting microscope and carefully studied the morphology of each specimen. "When I dissected the genitalia," she says, "the differences were pretty evident."

Eventually Calhau identified a handful of features that allowed her to differentiate between *G. mystaceus* and the large unidentified fly in the drawer. Later still, in other collections, she found more specimens of the new fly—some of them collected in Costa Rica, extending the range of *Gauromydas* species northward into Central America. She called the new species *Gauromydas papaveroi*, honoring Nelson Papavero for his contributions to Diptera taxonomy.[3]

A word about insect genitalia: When I spoke with Calhau I didn't fully appreciate just how important genitalia are to entomologists and insect systematists. In fact they are extremely complex, highly elaborate three-dimensional structures—particularly the penis. They evolve rapidly, in unpredictable ways, and are species-specific. And they provide taxonomists with an incredible amount of information. Two male insect specimens could appear almost identical in all their external characteristics—indistinguishable, even. A taxonomist who has spent a lifetime studying the genus might not be able to tell them apart. But if their genitals are different they can be considered separate species. The penis, or aedeagus, is one of the most important characteristics for identifying a new insect species. It might as well be a label. And it's true for all insects: beetles, weevils, damselflies, moths, and giant flies like *Gauromydas papaveroi*. Entomologists trying to revise a particular genus of insects might spend hours hunched over a microscope, trying to discern the subtlest structural differences between two penises.

But there's a problem: immediately after death, the penis begins to shrink. Its structure isn't preserved over time, so information is lost. All the unique and particular features an entomologist might use to identify a new species slowly become more and more difficult to detect. In recent decades scientists have developed better tools to preserve insect genitals—such as the Phalloblaster, a device designed by Australian entomologists in the 1990s to permanently inflate the penis so it forever maintains its shape. For a 2012 project, Colombian artist Maria Fernanda Cardoso even made large-scale anatomically correct models of insect penises—an exhibition she calls the Museum of Copulatory Organs.

There are long and multipronged penises and penises that spiral like a drill bit. A bean weevil (*Callosobruchus maculatus*) penis is terrifying, crowned with a bristling cluster of spikes like a medieval mace. Other insect penises are adorned with a ring of tentacles or topped with two long backward-curving horns, like fine-tipped hooks. A fruitfly (*Drosophila busckii*) penis is flowerlike in its intricate curling three-dimensional complexity. Some are nondescript, like blunted clubs or spindly sticks. A harvestman (*Thelbunus mirabilis*) penis resembles a deep-sea organism with a decorative fringe of whiplike extensions. In September 2016, near a mountain stream, Australian entomologists netted a novel species of crane fly (*Minipteryx robusta*) and later identified it by its strange and distinctive double-barreled penis.

Surprisingly enough, a few insect penises even just look like a dick.

Torsten Dikow is one of three curators in charge of the vast fly collection at the National Museum of Natural History in Washington, DC. Dikow has never seen a *Gauromydas* fly alive. If they seem rare, he says, it's because most of their life cycle—almost a year—is spent underground. Then, during their brief adult lives, lasting just a few weeks, they cruise through the air at high speed, near the treetops on the Brazilian savanna. Like the lightning cockroach, they are pretending to be something they are not. They have evolved, Dikow says softly in his melodic German accent, to mimic pompilid wasps or tarantula hawk wasps—large, powerful solitary wasps capable of killing tarantulas with their sting.

A tarantula hawk wasp is a creature straight from a nightmare.

The entire insect looks like a weapon, and it kills tarantulas. In 1990, when entomologist Justin Schmidt devised the Schmidt Sting Pain Index, a scale to rate the painfulness of different insect stings, he gave the tarantula hawk wasp a four, its highest rating. Nothing else comes close to it. The pain lasts only three minutes, but it warps time. Schmidt wrote, "Tarantula hawks produce large quantities of venom and their stings produce immediate, intense, excruciating short term pain in envenomed humans."[4]

Although harmless, *Gauromydas* flies are a convincing imitation. They impersonate the wasp, and they do it well. The similarity between a *Gauromydas* fly and a tarantula hawk wasp, which protects the flies from predation by birds in the wild, probably also prevents entomologists from collecting them in the field, Dikow says. Who would risk three minutes of excruciating pain to populate an entomology collection back home?

In São Paulo, Calhau turned her attention back to the specimens in the drawer. She selected two more and placed them side by side—a male and a female. The female was different from any other *Gauromydas* species she had seen before. Its abdomen—fat and sturdy and shaped like a bee's—was marked with a bright orange rectangle, oddly geometric against the black background. There were no labels on the specimen. Where the fly had come from, and when, was a mystery. A few months later Dikow was visiting the entomology collection at the Muséum National d'Histoire Naturelle in Paris—the third largest insect collection in the world. There, in a drawer filled with unsorted material, he found another specimen that bore the distinctive orange rectangle on its abdomen.

During his career, Dikow has described more than sixty new fly species. Often he finds unknown species by combing through unsorted material in old, established collections like the one in Paris. "When I visit another natural history museum or a university collection, that's the place that I go," says Dikow. "That's where there's something that nobody has really studied in detail. . . . Let's say a beetle researcher caught these flies: 'Okay, it's a fly. Put it over there.' Decades go by. Nobody ever looks at it."

Dikow arranged to borrow the specimen with the orange abdomen and sent it to Calhau. Collected in 1936, it became the holotype for another novel species, which she named *Gauromydas mateus*.

To identify an undetermined specimen, an entomologist must first be able to find it among the known and identified material. This is the battle for taxonomists. The Diptera collection Dikow curates in Washington, DC, contains more than three million pinned specimens. Even in a collection that size—drawer after drawer filled with hundreds or thousands of flies—there are just thirteen *Gauromydas* specimens. "If you're not looking for this genus, you're not going to see it," he says. "You're not going to go through those drawers."

Dikow says he's aware of at least two more undescribed *Gauromydas* species. In both instances, they are known from a single specimen—one distinctive oversize fly. One of them waits to be described in a drawer at the California Academy of Sciences in San Francisco. A single male specimen, it came from Trinidad—a region not known for *Gauromydas* flies. At the moment it is simply a large black fly with no name. It was collected in 1985. The decades slowly pass, and it remains unnamed. The other undescribed specimen comes from eastern Peru. These discoveries are important. Currently, so little is known about the flies, and there are so few known examples, that every single new specimen provides entomologists with a wealth of information about the evolutionary processes that produced them—a genus of superlatively large flies that expertly mimic a tarantula-killing wasp. There are still more unknown species to describe too, Dikow says—uncataloged in small, untended collections in Brazil. Almost no one is studying them.

In unvisited drawers, the big flies wait to be found, their wings outstretched forever as if in perpetual flight.

18

It Came from Area 51:
The Atomic Tarantula Spider
(*Aphonopelma atomicum*)

The Nevada Test Site is in the desert about sixty-five miles north-west of Las Vegas—a sun-scorched and arid place. Here and there, clusters of cubelike white buildings huddle on the red earth, bordered by an occasional airstrip or helipad. Arrow-straight roads crisscross the desert, heading nowhere. Here, beginning in January 1951, the United States Department of Energy detonated its nuclear test explosions, mostly underground—almost a thousand in all.

At the southern end of the test site, about twenty miles south-west of the mysterious facility known as Area 51, a flat, narrow valley runs north to south for seven miles. The ground there is still pockmarked, dotted with large rounded craters left by the atomic tests that took place sixty years before.

On June 29, 1961, at the height of the atomic age, arachnologist Willis Gertsch was collecting spiders at the test site. A few months later, in September, Operation Nougat began with the bomb test for Antler, a 2.6 kiloton underground detonation at the site. Gertsch set pitfall traps, digging deep holes between the craters and sinking cans into the soil, hoping spiders would fall in and become trapped at the bottom. First he captured a small black tarantula. Then another. Then a few more. That day Gertsch collected three females and fifteen males, placing them in vials filled with ethanol to preserve them. Later he deposited the specimens in the entomology collection at the American Museum of Natural History, where he was curator. In August 1961 Gertsch was at the test site again, amid the bowl-like craters, checking his traps and searching for the telltale signs of tarantulas: a deep circular burrow in the ground, like a hole made by a pencil, surrounded by a turreted mound of finely sorted soil. That day the result was a single male specimen.

The spider Gertsch collected is particularly distinctive because of its diminutive size. It's so small "it can sit pretty easily on a US quarter," says Chris Hamilton, an arachnologist at the Florida Museum of Natural History in Gainesville. Since 1961, Gertsch's vials of specimens have sat in the vast fluid collection at the American Museum of Natural History. In 2016, while revising the *Aphonopelma* genus of tarantulas for species found in the United States, Hamilton finally found them. He named the spider *Aphonopelma atomicum*, the atomic tarantula.[1]

A. atomicum is a miniature tarantula—one of eleven miniature species now known to exist within the *Aphonopelma* genus in the United States. Hamilton found the 1961 specimens during a research visit to the American Museum.

"It's a fascinating spider," he says. "The specimens collected during that time period at the Nevada site were all within two larger jars." The same collection includes a handful of even older *A. atomicum* specimens: three males collected in October 1959 by an unknown collector. Most likely, he says, the specimens from 1959 and 1961 were the first examples of miniature tarantulas ever collected. Before that no one knew they existed. The insect's small black body measures about six millimeters across. Its egg-shaped abdomen is covered with a down of black hair. It's a perfectly miniaturized version of the more familiar hand-size tarantula species.

Working carefully, Hamilton pulled the specimens from the vials—a profusion of wet black legs covered in urticating hairs. Unlike many insects, which have durable exoskeletons, tarantulas often are preserved in ethanol. This allows arachnologists to manipulate still-flexible specimens decades after they were collected, investigating aspects of their morphology that would be lost forever in dried specimens. In all, Hamilton examined three thousand specimens—many of them borrowed from the American Museum collection.

"A huge amount of the tarantula diversity [at the museum] is stuff that hasn't been identified yet," he says. Curator Louis Sorkin sent me a photo of part of the collection: steel cabinets that house wooden boxes filled with flasks. One of them is simply labeled '*Aphonopelma* sp. UNDETERMINED.' Hamilton gathered the material together and returned to Auburn University. "I took hundreds of specimens back with me," he says.

There are only a handful of known *A. atomicum* specimens, all from the same limited region of the Southwest: the low mountains and foothills along the California-Nevada border that enclose Death Valley and the Armagosa Desert.

For several decades in the mid-twentieth century, Willis Gertsch was the most prominent arachnologist in the United States. He was a prolific collector and taxonomist. By the time he died in 1998, aged ninety-two, he had described more than a thousand new species. In 1949 he published a popular science book titled *American Spiders.*[2]

A. atomicum represents a taxonomic mystery to Hamilton: Why didn't Gertsch describe the spider after he collected the holotype in Nevada in 1961? It was small, unusual, and very distinctive. "He obviously saw them," says Hamilton. "I don't really understand why he didn't describe them, because they're radically different, particularly at that time, from anything else he would have seen tarantulawise in the United States."

Erected in 1901 by prominent English arachnologist Reginald Pocock, the *Aphonopelma* genus consists of about ninety tarantula species native to the Americas. In the United States they are found west of the Mississippi River, across the Southwest to California. Mostly they are found in the hot and arid places—in the Mojave Desert, in Death Valley, among the saguaro cactus in the Sonoran

Desert, in the Chihuahuan Desert at the base of the Chiricahua Mountains in Arizona. They live on the outskirts of small half-forgotten boomtowns with names like Apache and Tombstone, where coyotes prowl at night and tumbleweeds bounce along wind-blown streets. Across the southwestern United States, tarantulas roam the cracked sunbaked parking lots and the dry thornscrub.

But they live elsewhere too: in the rolling foothills of the Sierra Madre in California and on the flat green prairies of southern Colorado. They are found in Louisiana and Missouri and Arkansas. Their range extends southward through Mexico and Central America. For practical reasons, Hamilton confined his project to the United States.

"At the beginning of the project there were fifty-five described *Aphonopelma* species in the United States," says Hamilton. But more than half of them were synonymies—a single species named twice or three times. In those cases the earlier species name stands: the principle of priority. "If you're calling something two different species and they're not, it can have dramatic outcomes," he says. "You want to get that right."

For instance, *A. behlei* was named in 1940 by Ralph V. Chamberlin, a tarantula systematist who completed a lot of the early work on the *Aphonopelma* genus. But, Hamilton says, it's actually *A. marxi*, a species that already had been named in 1891 by Eugene Simon. Chamberlin also named *A. coloradanum*, *A. echinum*, *A. harlingenum*, *A. waconum*, and *A. wichitanum*. But they're all *A. hentzi*, a common large, light brown species first described in 1852 by Charles Girard. In other words, he named the same species five times. Another common species—*A. iodius*—was first named in 1939 by Chamberlin and Wilton Ivie, then renamed again and again: *A. brunnius* (1940), *A. chamberlini* (1995), *A. iviei* (1995), *A. lithodomum* (1940), *A. smithi* (1995), and *A. zionis* (1940).

Chamberlin was responsible for three of those later names too. Hamilton abolished all the synonyms. There are now twenty-nine species in the United States. Fifty-five previously named species were reduced to fifteen, and fourteen new species were added.[3] They exist in five distinct lineages, he says. Several other species were classified as *nomina dubia*, doubtful names. They probably don't exist at all. Many were named from a single specimen, collected once and never seen again—seeds of error planted mostly in the 1930s and 1940s.

Several of the new species waited decades in museum collections to be described: *A. chiricahua* was first collected in 1956 and is endemic to the Chiricahua Mountains in southeastern Arizona. It lives at high elevations, among the pine-oaks on sky islands—verdant, steep mountains surrounded on all sides by inhospitable desert. This type of geography produces isolated ecosystems with their own endemic species: the Chiricahua fox squirrel (*Sciurus nayaritensis chiricahuae*), several frogs, more than sixty snail species, and numerous plant species are found there and nowhere else.

Almost nothing is known about the natural history of *A. chiricahua*. Three other new species—*A. parvum*, *A. prenticei*, and *A. xwalxwal*—were collected in 1963 from sites across the Southwest. One of the type specimens of another new species, *A. saguaro*, had no date on it, but it was clearly old, Hamilton says. It was collected by Owen Bryant in Saguaro National Park in southern Arizona. A prolific collector, when Bryant died in 1958 he left his entire estate to the California Academy of Sciences. He had collected botanical specimens from the same region in 1936; the spider was likely collected about then. Hamilton named another species *A. johnnycashi*. The holotype—impossibly large, black and long-legged—was collected near Folsom Prison, made famous by Johnny Cash in the song "Folsom Prison Blues." Eight of the newly described spiders are miniature species, representing the majority of the new diversity.

With his hair slicked back, Hamilton looks like he stepped out of the rockabilly 1950s. His arms are tattooed: on one forearm, *Darwin*; on the other, *Wallace*. He jokes that he can show the tattoos to creationists if a fight breaks out. Other tattoos depict a snake skull, a spider, his hero Johnny Cash, and Darwin's evolutionary tree of relatedness—a sketch in one of Darwin's journals that is the first image of a phylogenetic tree. By nature Hamilton is quiet. When he speaks he pauses often, struggling to find the right words to make a point. But he considers himself a storyteller. Before he became an evolutionary biologist, Hamilton was a freelance photojournalist working in war-torn places—Sierra Leone during the armed conflict in 2000, Liberia in 2005.

Now he tells profoundly interesting stories about tarantulas, using the specimens as his medium. And he is revolutionizing what we understand about their distribution, their evolution, and their gradual dispersal across the southern United States. For this, the collections are instrumental. At times Hamilton has collected fresh specimens too—about 1,800 of them. They came from across the southwestern and the western states and via a citizen-based science program run by the American Tarantula Society. Fresh specimens are important, he says. "It's very difficult to grab a specimen out of a jar and identify it to species correctly," he says. "Things just sit in ethanol. They fade. If there was any type of color that may have been helpful, it's gone."

For a decade he and his coauthors crisscrossed the United States collecting tarantulas—from the shoulder of desert highways, prairies, and forest trails. They came from parking lots, the front steps of schools, campgrounds, and miniature golf courses. "You take all your field equipment and all your camping gear, pack up the vehicle and just drive around," says Hamilton, who hunts specimens for weeks at a time. "Each day you're at a different spot, sometimes multiple spots in a day depending on how the collecting goes; camping in parks or on open land, wherever, at night, then moving on the next day and just trying to cover as much ground as you can."

He pours water down a burrow and, hunched over it, waits for the spider to emerge, its legs appearing first. Or he digs it out, working slowly with a trowel to loosen the hard earth and enlarge the burrow. At night, during the mating season, he finds males walking boldly across western streets in the darkness—many-legged shadows looming across a silver road in the moonlight.

To determine the boundaries of each species, Hamilton measured thousands of specimens: the width and length of different legs and of the carapace, and certain features of the genitalia, which show significant variation between species. He mapped out via GPS where each specimen had been collected and prepared precise maps showing the limited distributions of each species. Additionally, he extracted DNA from the fresh specimens and amplified it using next-generation sequencing. Then he compared the genomes of spiders, detecting small but conserved differences between species.

"The advantage they have is using DNA," says London-based tarantula expert Andrew Smith, author of *Tarantula Spiders: Tarantulas of the USA and Mexico* (1994).[4] He spent much of the 1980s in the American Southwest collecting tarantula specimens, many of which are now in the collection at the Natural History Museum in London. Back then, Smith relied solely on morphology. The molecular technology that allows Hamilton to compare the DNA of different species wasn't available then; it changes everything. A few decades ago Smith described several of the species that Hamilton has subsequently identified as *nomina dubia*. Specimens like *A. atomicum* sit unidentified in collections for long periods, Smith says, because there are too few specialists to identify them.

"I basically popped up in the mid-1980s," he says, "and nobody had worked on this stuff since the 1930s. The museums had time to accumulate new material." Often the new material is identified as a common species like *A. iodius*, he says. "There may be 120 specimens in a jar, and twenty, thirty, or forty jars, all labeled the same," he says, chuckling. "Unless a researcher has got an enormous amount of time to go through all of that material, checking each specimen in turn, it's simply going to be disguised."

Smith considers himself an arachnohistorian. He has recently returned from Brazil, where he shot a documentary called *1640: The First Tarantula Spider*. "There are three paintings, painted between 1640 and 1648," he says. "They're Dutch and from a period of almost-forgotten history when Holland, for about forty years, controlled Brazil, based around Recife. In that very short time, the Dutch being the Dutch, they sent in scientific expeditions, and a huge amount of material went back to Holland. They had naturalists, painters—all basically collecting data." Each of the three paintings includes a tarantula. "We've actually returned back to the Recife area and tried to find the spiders in the paintings."

In 1636 the German naturalist Georg Markgraf accompanied

one of the first Dutch expeditions to Brazil, collecting numerous specimens from the region—botanical and animal. Dutch artist Albert Eckhout was there too, documenting the expedition via a painted account. "There's a famous painting by Eckhout of a native Indian," says Smith. "Tucked in the left-hand corner of the painting you'll see a tarantula spider. When you look at it, if you're a tarantula buff, it's immediately obvious it's *Theraphosa blondi*. But that spider can't be found in the Recife area." It presents a conundrum. Smith thinks the spider came from farther away—near a Dutch fort to the north, deep in the Amazon basin. "I suspect that Markgraf brought that spider back to Recife saying, 'Look at this thing! You've never seen a spider this size.' *Theraphosa blondi* is the world's biggest tarantula. It's a beast—the size of a dinner plate. You can imagine a Dutch naturalist of that period—early seventeenth century. He pins down the world's biggest spider. You're going to bring it back and show everybody, aren't you?" Then it was inserted incongruously into the painting. "The artist of course has one look at this spider, and any plan he has to put a local spider in the corner of his painting has just gone out the window," Smith says. "He's now popped this thing in, because it's a beast and he loves it. And he's an artist."

Aphonopelma catalina is known from just six specimens collected in and around the Santa Catalina Mountains near Tucson, Arizona. Perhaps elsewhere in the western and southwestern United States, unknown *Aphonopelma* species are still waiting to be discovered. They might live in limited microenvironments that haven't yet been surveyed by arachnologists: a sunbaked arroyo in New Mexico, the outskirts of a built-up city like Los Angeles or Las Vegas, or a prairie in Colorado. Hamilton believes there likely are more unknown species to discover on the isolated sky islands in southwestern New Mexico and southeastern Arizona—a challenging place to survey, he says. Unknown specimens from those regions could already be sitting in ethanol in natural history collections. "Sometimes you end up with specimens getting stacked in jars and pushed way to the back," says Smith.

Like *A. saguaro*, collected sometime in the 1930s from between the towering straight-trunked saguaro cacti, perhaps they were collected once before, retrieved from a pitfall trap a hundred years ago or longer—mislabeled or unnamed—and placed in a flask of ethanol. And in the future, he says hopefully, perhaps some of his *nomina dubia* species will be reinstated too. All it takes is the unexpected discovery of another specimen buried beneath a hundred *A. iodius* specimens in a jar, and the species he once described can become real again. In the meantime the flasks wait on shelves and in cabinets, brimming with possible meaning.

19

The Host with the Most:
The Nematode Worm
(*Ohbayashinema aspeira*)

A long time ago in the American West, ringed by the mythic peaks of the Teton Range, ecologist Olaus Murie bent to retrieve a dead American pika from the ground. Found at higher elevations across the western states, the pika—its scientific name is *Ochotona princeps*—is a small-bodied member of the rabbit family with rounded, mouselike ears and shiny eyes that gleam like little black buttons. But this particular pika specimen, Murie soon discovered, was more than it seemed. It was a host, too. It contained a multitude of parasites that twirled and ribboned silently through its intestines.

Eric Hoberg is a research zoologist for the USDA's Agricultural Research Service and an expert in parasites—particularly helminthic worms. He is a curator of parasitic nematodes at the National Museum of Natural History in Washington, DC. In 2010 Hoberg described a novel species of nematode worm that was harvested from the pika Murie collected in 1930. It had waited eighty years to be named.

The worm—"it looks like a little corkscrew," says Hoberg—was baptized *Ohbayashinema aspeira* in a paper in the *Journal of Parasitology*.[1] There is a photo of Hoberg taken about the same time he described the worm. In it Hoberg, who is in his sixties and gray-bearded, holds a tall glass flask filled with what looks like a half-pound of freshly cooked noodles. The caption reads: "Eric Hoberg, chief curator of the U.S. National Parasite Collection, examines roundworm specimens from raccoons, which pose a potential threat for human infection."

Specifically, the roundworms in the flask are *Ascaris columnaris*—collected from a raccoon's small intestine near Harrisburg, Pennsylvania, in December 1930. But all around Hoberg, the shelves are lined with jars filled with archived parasite specimens, some many decades old, some even centuries old. The labels are handwritten and faded. "It is arguably among the largest collections in the world for helminth parasites, worms primarily," he says. All together, he estimates, it contains about twenty million specimens.

The collection represents the inhabitants of an unseen world. Immediately next to him on the shelves are nematodes from the stomach of a pygmy sperm whale (*Kogia breviceps*); intestinal parasites collected from sheep in Quincy, Florida, in 1934, floating in their flask like a slowly unspooling ball of yarn; and a jar of ghostly liver flukes—each thumb-sized—harvested from cattle in 1921 in Fort Worth, Texas. It's all here.

In other words, Hoberg is the keeper of a house of horrors.

Murie collected the pika in Wyoming on October 30, 1930. Opening up the animal, he removed the entire gastrointestinal tract, found the tenants waiting inside, and sent them to the National Parasite Collection.

By December 1930 Gerard Dikmans, a nematologist at the USDA, had described a worm harvested from the pika, a relatively long, straight-bodied nematode measuring ten to fifteen millimeters. Dikmans named it after Murie: *Murielus harpespiculus*. But there was another worm in the vial that Dikmans failed to see because it was obscured by the larger and more conspicuous species. After Dikmans described the larger worm, the vial containing the type specimens was accessioned into the National Parasite Collection. There it was placed on a shelf alongside other flasks filled with liver flukes and pinworms and roundworms from raccoon intestines. And then it was mostly forgotten.

Northern Michigan University biologist Kurt Galbreath was the first to notice *Ohbayashinema aspeira*. During fieldwork in Wyoming and Washington State, Galbreath—one of Hoberg's coauthors—had been trapping pikas and harvesting their intestinal parasites, just as Murie had done in 1930. Among the *M. harpespiculus*—a parasite he'd expected to find there—Galbreath saw something else.

"There were some small oddball nematodes that were only found at one locality he collected from," says Hoberg. "We didn't know exactly what they were. We kind of put them aside for the time being." To investigate further, Hoberg decided to retrieve the eighty-year-old type specimens of *M. harpespiculus* from the National Parasite Collection. "I just poured out the vial into a dish and thought: Oh, there's something else here." They were the same worms that had confused Hoberg and Galbreath in the more recent samples: the oddballs—minuscule, and spiraled like corkscrews. There were just two of them in the vial, both female. At first they were almost too small to see at all, floating in the dish like tiny coils. "It turned out to be the discovery of these specimens that had been sitting since 1930, unrecognized."

The Nemata phylum is incredibly diverse and includes at least twenty-five thousand described species. Many of them are free-living. Some, known as helminths, are parasitic—they live inside other species. Hoberg estimates there might be as many as a million nematode species in total, with most of them still undescribed. They exist everywhere, in all environments. In the brittle, almost lifeless Antarctic landscape, *Scottnema lindsayae*, a millimeter-long nematode, is one of the most widespread land animals. It thrives on a diet of algae and bacteria. During times of hardship, when water is scarce, it slows its metabolism and desiccates, waiting in the dry, rocky soil of the Antarctic valleys for water to return. Other nematodes live in deserts or in the briny mud on the ocean floor, or in volcanoes. *Panagrellus redivivus*, known commonly as the German beer mat nematode, is a free-living microworm found almost exclusively on the absorbent mats on top of German bars. Helminthic parasites inhabit everything on Earth, too: mammals, birds, insects, fish. Growing to more than twenty-eight feet, *Placentonema gigantissima* parasitizes the placenta of sperm whales. *Mermis nigrescens* targets grasshoppers, unspooling extravagantly to almost eight inches long, a squirming coil that emerges from a dead grasshopper like a magic trick no one wants to see. Even earthworms (*Lumbricus terrestris*) are parasitized by nematodes — worms within worms.

In 1914 Nathan Augustus Cobb, an American nematologist who laid the foundations of nematode taxonomy, wrote,

In short, if all the matter in the universe except the nematodes were swept away, our world would still be dimly recognizable, and if, as disembodied spirits, we could then investigate it, we should find its mountains, hills, vales, rivers, lakes, and oceans represented by a film of nematodes. The location of towns would be decipherable, since for every massing of human beings there would be a corresponding massing of certain nematodes. Trees would still stand in ghostly rows representing our streets and highways. The location of the various plants and animals would still be decipherable, and, had we sufficient knowledge, in many cases even their species could be determined by an examination of their erstwhile nematode parasites.[2]

Nematodes live within a protective outer cuticle. "In this group of nematodes there are structures on the surface cuticle that are diagnostic," Hoberg says. Using a microscope, he carefully characterizes a worm, counting the ridges that corrugate the cuticle and closely examining their structural subtleties. This worm is unusual: it coils tightly around itself — ten to twelve spirals in total except for the anterior part of the body, which is straight, hence its species name *aspeira*, meaning "spire" in Greek.

It's July 2015. Hoberg is seated in the cramped cabin of a single-engine float plane, flying across the Northwest Passage to Cambridge Bay, a hamlet of 1,500 people on Victoria Island, in Nunavut, Canada. He is in the Far North, above the Arctic Circle. The native people in Nunavut speak Innuinnaqtun. Hoberg will work with them to obtain samples from the large mammals they hunt: elk, moose, and caribou. The tundra rolls slowly underneath Hoberg—a wide, flat, endless expanse of green land interspersed everywhere with small, irregular bodies of water. Behind him in the cabin, pinned to cardboard, are several plump vole and lemming specimens collected from the tundra below.

Hoberg will harvest the parasites from their intestines too. Eventually the specimens will provide a unique window, allowing him to better understand the complex Arctic ecosystems passing below him. Since the 1970s, with a team of collaborators, Hoberg has been conducting a long-term biotic inventory of mammals and their pathogens across the Arctic. His work has taken him to numerous remote, snow-choked places on the rim of the world: to Siberia, to the Aleutian Islands, and to many sites across Alaska and the Canadian Northwest Territories. He has been on extended field expeditions to Finnish Lapland, to Talan Island in the Sea of Okhotsk, and to Nunavut—the northernmost and least populated territory of Canada. He has spent time at Palmer Station, a US Antarctic Program research base on Anvers Island, off the Antarctic Peninsula. In photos, Palmer Station looks like a few antenna-topped sheds perched on a rocky barren shelf and backed by a wall of ice cliffs. Together the team explores patterns in the diversity and distribution of different parasites in Arctic mammals. "Everything from the tiniest shrews to the large ungulates, such as moose, musk-oxen, and caribou, and the carnivores," Hoberg says.

In other words, a worm is more than just a worm. Harvested from the anterior small intestine of a small mammal—from a pika in Wyoming or a lemming in Nunavut—a single worm can provide Hoberg with a wealth of information about the animal it came from. But it can tell him more. It can also inform him on the larger population it belongs to, and the way that population interacts with diverse species in a much larger ecosystem. A tiny parasite hidden in the small intestine of a mammal can reflect much larger changes taking place.

Across the Arctic, the environment is changing. Global sea ice sheets are retreating, more quickly than anticipated. Temperatures are rising. Ecosystems respond. The pikas move to new landscapes. The ungulates change their habits. Carnivores begin to hunt new prey. Sometimes a parasite will find a new host. Occasionally host populations collapse. The entire complex chain reaction—a story that involves multiple players moving in and out of different habitats—can be revealed by the presence or absence of a parasite in the gastrointestinal tract of a pika, or an elk, or a malnourished Arctic fox.

Hoberg considers himself a biogeographer, not a taxonomist. His taxonomic efforts, although important to the field of parasitology, are incidental, he says. He's pragmatic: he's named species because they had no names. The newly described species provide him with a fuller understanding of the gathering changes he sees taking place within a fragile ecosystem. His early efforts at parasitology—first at the University of Alaska and then the University of Saskatchewan in the 1970s—concentrated on populations of marine birds in the North Pacific and Bering Sea.

"My interests in parasitology are focused on what stories parasites can tell us about the larger biosphere," he says. Research that began with marine birds has expanded over four decades to include the entire biodiversity of fauna across the Holarctic—a vast global faunal region that encompasses the northern latitudes worldwide.

All together, Hoberg has described forty-two new species, twenty new genera, and one new order of parasitic worms. He's currently in the middle of characterizing several new species of tapeworm—he calls them "tapes" for short—from numerous host species. Museum collections are vital to his taxonomic work, he says. They provide a temporal record of host parasite relations that can't be accessed any other way—a record across time. The results of biotic surveys are like a snapshot of a moment in time; the differences between an Arctic biotic survey conducted in the 1970s and another in 2017 are meaningful. To Hoberg, comparing the results of parasite surveys across time is as useful as time-lapse images of the Arctic ice sheets shrinking over time are to me.

"I was just handling a specimen this morning that was 210 years old," he says matter-of-factly. "It's a chunk of a tapeworm that was collected by Rudolphi just shortly before 1810." Karl Rudolphi, a Swedish naturalist, died in 1832, but not before earning a title no one else wanted: the father of helminthology. The tapeworm belongs to a species known as *Moniezia expansa*, Hoberg says. "It represents the eighteenth specimen that was accessioned in the US National Parasite Collection."

Found in ruminants, *M. expansa* is a superlative worm. It can grow up to ten meters long—a flattened, endless pale ribbon, growing unseen in the intestinal tract. Hoberg has retrieved the specimen from the collection for a colleague—Voitto Haukisalmi at the Finnish Museum of Natural History, who is researching tapeworm diversity in ungulates like reindeer and moose. Hoberg suspects that *M. expansa* is not one species at all, but a complex of many closely related species.

The world is changing, Hoberg says. Every time he returns to the Arctic, flying over the tundra of the Northwest Territories along the indented Bering Sea coastline, he sees the changes firsthand. Intestinal parasites help researchers like Hoberg understand the subtle impacts some of those environmental changes have on fragile ecosystems. "Look out your window: you see the biosphere as it is," Hoberg says. By using natural history collections, he can see the biosphere as it once was. The archives only gain power over time. "They're basically the foundations for everything we understand about the biosphere."

From a Time Machine on Cromwell Road: Ablett's Land Snail (*Pseudopomatias abletti*)

Barna Páll-Gergely removes a slender glass vial about the size of my index finger from a small, sturdy box. Inside is a delicate spiraled shell, tapered like a turret, kept in place by a cocoon of protective archival polymer wool. The shell is the color of bone, ribbed along its surface.

Empty now, the shell once belonged to a Himalayan land snail. Although it was collected more than 150 years ago, it remains unnamed. With the years, it has become an artifact. It's one of about nine million specimens in the mollusk collection at the Natural History Museum in London. The oldest of them were bequeathed to the British Museum by Hans Sloane, who died in 1753, and were among the earliest objects at the museum. The collection fills the drawers and cabinets around Páll-Gergely—a city of drawers and little boxes. All together it contains about sixty-six thousand type specimens—voucher specimens for entire species.

Páll-Gergely, a Hungarian postdoctoral fellow at Shinshu University in Matsumoto, Japan, is a malacologist: he studies shells. He carefully places the shell beneath the objective lenses of a dissecting microscope and leans in to examine its whorls and sutures under magnification. "I work on the classic taxonomy of different kinds of land snails, mainly southeast Asian groups," he says.

In March 2015 Páll-Gergely published a revision of the *Pseudo-pomatias* genus of land snails in the journal *Zootaxa*.[1] The genus is found from the Himalayas to Taiwan. It comprises relatively small gastropods with turriform shells five to fifteen millimeters in height. They are slender and high-spired. Taxonomically they belong to the Pupinidae family of operculate land snails. "They have a little door called an operculum," he explains. "When they go inside the shell, the door seals the aperture."

The small, tapered Himalayan shell beneath the microscope is one of them—a *Pseudopomatias*. Páll-Gergely has named it *Pseudo-pomatias abletti* after Jon Ablett, the curator who has overseen the mollusk collection at the Natural History Museum for more than twelve years. A small, stubby shell, *P. abletti* is about six millimeters in height, squat and round-bellied. The holotype was collected in late 1859 by the English naturalist and malacologist William Thomas Blanford. It came from Darjeeling in northern India, at the green feet of the Himalayas. It closely resembles another common species found across the region, *Pseudopomatias himalayae*. *P. hima-layae* also has a pale, turreted shell, with dark gray sutures tracing a spiral as fine as a pencil line all the way to its blunted tip, known as the protoconch. But there are subtle differences that set it apart. When Páll-Gergely peers into its unusual, slightly triangular, aperture and studies the sutures and the ribs that line its whorls, he immediately sees its defining traits.

Páll-Gergely found the holotype of *P. abletti* in the mollusk collection in London, in a box that supposedly contained only *P. hima-layae* specimens. On top of the box is a label handwritten in thick black ink: "Pomatias, Himalaya, Darjiling, W. T. Blanford, Esq." Inside the box were four shells—shiny little cream-colored spires—but only two of them are actually *P. himalayae*, he tells me. The other two shells belong to the species he subsequently named *P. abletti*. They're different, but not quite different enough to be noticed by a nonspecialist. They went unnoticed for more than 150 years. The same situation is true in collections across the world, he says: intermixed with the *P. himalayae* shells are the smaller *P. abletti*, with their slightly flattened whorls. "The people who were looking at these materials didn't recognize that there were two species within one sample."

When Páll-Gergely surveys the drawers at the Natural History Museum for specimens of Himalayan and Southeast Asian species, he is looking backward through time—collaborating with some of the founders of modern malacology. To make his comparisons, he examines the type specimens that in some cases were collected by malacologists working more than two centuries ago. Thousands of the small spindle-shaped shells in front of him were collected by a handful of intrepid Victorian English scientists. Together they formed a tight-knit network of collectors and taxonomists. They were hunting for tropical and subtropical land snails—gastropods, not tigers—but their work was dangerous and took them to some of the most inaccessible places on Earth.

Páll-Gergely spent weeks in the archives overseen by Ablett in London, measuring specimens collected by Henry Haversham Godwin-Austen. Born in 1834, Godwin-Austen was an explorer, cartographer, and amateur naturalist. He worked for the Great Trigonometrical Survey of India, an effort by the East India Company to map remote and uncharted regions of the country, including the Himalayas. Since 1802, the survey had been demarcating the extent of the British territories and measuring the heights of several of the tallest mountains on Earth-many of them tower eight miles into the air. Godwin-Austen joined the effort in 1856 as a young man. Previously he'd helped to map the Irrawaddy River Delta in Myanmar. Shortly after arriving in India he'd been attacked and beaten, and he'd had to return to England for a year to recover. But by 1860 he was back in Kashmir, surveying the Lesser Karakorams, a vast complex of windblown ranges with the greatest concentration of high mountains in the world, including the Himalayas. There in the mountains, at high altitude, Godwin-Austen, bearded like a schnauzer, was an indomitable presence—some called him the greatest mountaineer of his time.

At 28,251 feet above sea level, K2 is the second highest peak in the world after Mount Everest. Officially it's named Mount Godwin-Austen. Its peak, like a massive snow-clad inverted V, rises from the base of the Godwin-Austen glacier. To pinpoint the exact location of its elusive peak—and confirm that it lay within the British Empire—Godwin-Austen climbed through a vertiginous field of ice-clad peaks into the sky. Making his way along a windswept spur of the Masherbrum (K1, 25,659 feet), he ascended another two thousand feet, dangling above the Muztagh Glacier to fix the position of K2. Despite all his high-altitude achievements, Godwin-Austen was most enthusiastic about small Indian land snails—carefully sifting through the leaf litter at the base of towering mountains like a prospector to collect their vacated shells.[2]

Like almost all Victorian malacologists, in the field Godwin-Austen used native collectors: "I found them very keen of sight, and they very soon knew the different genera and those I most wanted," he wrote in 1892. "Mr Doherty was far too liberal in giving the Nagas two rupees for shells. In the forests and noisome jungly ravines, at times swarming with leeches, this work is not over pleasant, and it is only interest and excitement of finding some beautiful new form at any moment that leads one on this lowly chase, as it has me for many an hour."

Blanford—who first collected *P. abletti* in Darjeeling—worked as a civil servant too. For almost thirty years, beginning in 1855, he was a geologist for the Geological Survey of India. His brother worked alongside him. Everywhere they went they scoured the wet undergrowth for snails, sending their shells to collaborators across the world. In January 1859 William Henry Benson made the first description of *P. himalayae* from the shells Blanford had sent him. It was published in the *Annals and Magazine of Natural History*. "A young shell," he wrote, "is of a clear pale horn-colour." The morphological description, in Latin, sounds like an incantation: "Testa perforata, attenuato-turrita, solidiuscula, oblique confertim crassicostata."

Benson had worked in India for the East India Company. He described several new mollusk species from across Asia. On his death in 1870 he bequeathed his collection of thousands of shells to Sylvanus Hanley, a conchologist and malacologist known for publishing the first scientific study of shells that used photography.[3] Famously, Hanley removed all the locality data on Benson's type specimens, replacing it simply with "India." Together Godwin-Austen, Blanford, Benson, and others laid the foundations of modern malacology.

Páll-Gergely is bespectacled and soft-spoken. Since 2011 he has lived and worked in Japan—at Shinshu University in Nagano Prefecture. He has been fascinated with shells, he says in his perfect, carefully chosen English, since he was a child growing up in Hungary. By the time he was five years old he was already collecting big, colorful specimens of marine snails. As a teenager he became interested in land snails and began amassing his own collection—shells from the Romanian Carpathians, southern Europe, and Turkey. He began studying Asian land snails when working on material collected by his friend András Hunyadi, an amateur Hungarian malacologist. In total Páll-Gergely has described almost forty taxa from Hunyadi's collections alone. "I realized that finding species which have not be seen by anybody and reclassifying known ones based on novel morphological characters is really fascinating," he says.

Páll-Gergely borrowed shells from elsewhere, too: from the Senckenberg Museum in Frankfurt and the enormous old collection at the National Museum of Natural History in Paris. A lot of the shells in European collections are old and undated; the material has begun to show its age. "In the Vienna Museum in Austria," he says, "in many cases, the century-old samples are completely destroyed."

Surprisingly, archived shells are no more durable than dried mammal skins or snakes kept in fluid. Shells stored in collections for a long time often begin to exhibit Bynesian decay, a chemical reaction between the acidic fumes that build up over time in wooden cabinets and the calcium carbonate surface of the shell, which is basic. The damage is irreversible. The surfaces of old mollusk specimens become pocked and lined with salt crystals that resemble interconnected clusters of mold.

In addition to morphology, Páll-Gergely extracted DNA from some specimens, sequencing it to test the phylogenetic similarity between *Pseudopomatias* shells and groups of other terrestrial snails. Although the shells are essentially vacated living spaces, the snail leaves traces of its DNA on the whorls and spirals inside the shell. He carefully makes a tiny hole in an upper whorl and pipets digestion buffer into the shell to obtain the DNA. From that he can generate molecular data, directly comparing the genomes of different species. All together, Páll-Gergely described eleven new species in the revision. "Maybe eight of those were from very old museum collections," he says.

Originally they came from across Southeast Asia and remote northern India and from the hilly northeastern part of Vietnam that extends along its border with China. Some were collected from Sikkim, a landlocked Indian state nestled between Tibet to the north, Bhutan to the east, and Nepal to the west. Others were found in the wet undergrowth on the lower elevations of the Himalayas—from remote overgrown, half-forgotten places backed by an interlocking wall of peaks—and from farther east, along the rumpled green border between Myanmar and India.

"Generally speaking, because the material of the shell is calcium carbonate, they prefer limestone habitats," Páll-Gergely says. "There are exceptions, but you can find the largest diversity of land snails around limestone mountains." This makes the shells relatively easy to collect. "If there are big vertical limestone cliffs, you just have to collect the material, the leaf litter for example, from the bottom of the rocks. That way you can have access to a lot of shells."

In most instances the novel species had been misidentified. "In some cases there is one name on the label, but there are two or even three species in the glass vial. It's relatively easy to see the differences if they are all from the same lot," says Páll-Gergely. In other cases collectors knew the moment they found the specimens that they were unnamed species, but for one reason or another they didn't formally describe them. "They recognized that it was a new species," he says. "They even gave it a name, usually just according to the most obvious shell character, or from the locality, and they just wrote it on the label. But they left it like that. It wasn't scientifically known."

P. shanensis is described from a single shell collected from a remote location in Myanmar in 1895. Páll-Gergely didn't see it in any other collection. *P. reischuetzi* is known only from the type series, a handful of shells Godwin-Austen collected at Lhota Naga in 1903. Godwin-Austen collected another new species—named *P. harli*—from remote western Bhutan.

The shells are important enough by themselves as single, irreducible components of biodiversity. But the collections they are part of are even more important because of what else they can tell us. Like the African dragonflies identified by Klaas-Douwe Dijkstra, snails are sentinels that inform us about the relative health of complex ecosystems. According to the International Union for Conservation of Nature, mollusks are among the most affected of all phyla by increased rates of extinction. For a start, they can't outrun anything. As local conditions change, the snails must either adapt or disappear. They are barometers of change. Many of them disappear. Most likely we are underestimating mollusk extinctions: no one notices when an unnamed species disappears. But eventually natural history specimens allow malacologists to detect broad, slow environmental changes: adaptation, decline, invasion, extinction. Research has shown that snails in regions affected by climate change adapt by producing lighter-colored shells to reflect more sunlight. But other species are disappearing altogether. Across the Pacific region, numerous snail species have been replaced by a single invasive species, *Euglandina rose*, a predatory snail. An old collection like the one Ablett oversees is a time machine housed in a massive stone cathedral-shaped edifice that rises into the gray air on the Cromwell Road in West London. These changes might take place over a hundred years or more, but they're all reflected in the collections, allowing a researcher like Páll-Gergely to step through a portal and arrive in Darjeeling in the 1860s, or Bhutan, or southern China, or anywhere else at any other time he chooses to travel.

There are still many other shells, he says, waiting for someone to find them and recognize something unknown: the precise curve of the ribbed whorls, or the shape of the aperture, with its rim shaped like the flared bell of a trumpet, or the pinna of an ear. One by one he is finding them and giving each a name.

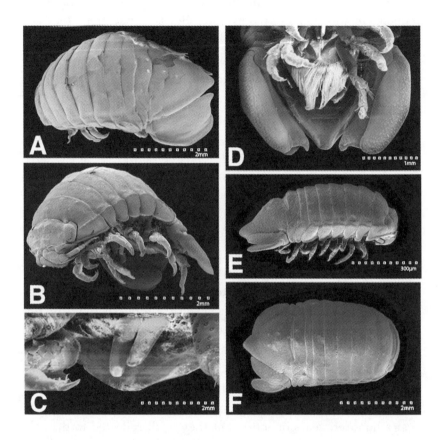

In Sight of Land:
Payden's Isopod
(*Exosphaeroma paydenae*)

Under a scanning electron microscope, *Exosphaeroma paydenae* resembles an enormous monochrome armadillo. Its smooth, curled body is covered with rows of bony protective plates—a rigid exoskeleton. In fact, although on a monitor *E. paydenae* looms like a ghostly armored vehicle, in life it's smaller than a pea. At its largest, it's less than seven millimeters long. It's an intertidal isopod, known colloquially as a marine roly-poly, the aquatic equivalent of a terrestrial woodlouse. The holotype of *E. paydenae* was collected in 1873. Unexamined for more than a century, it was finally described in 2015 by taxonomists at the Natural History Museum of Los Angeles County.

One day in March 2004, Dean Pentcheff, a long-haired, gray-bearded research associate at the Natural History Museum of Los Angeles County, was wading in the shallow intertidal water of the Pacific Ocean, near the southernmost tip of Los Angeles. Scattered along the wrack line were students from Loyola Marymount University, wearing boots. They'd taken a field trip with Pentcheff to the narrow shoreline at Paseo del Mar for an introductory course in invertebrate zoology. Beyond the rocky beach, the working-class neighborhoods of San Pedro extend inland. In the distance to the north begins the endless sprawl of the city of Los Angeles. Less than a mile to the west the cranes of the Port of Los Angeles pivot and lift, tirelessly stacking shipping containers.

Pentcheff had found something in the water. He gently pulled a common sea star to the surface, dripping seawater. Attached to one of its five arms was a marine roly-poly—small and armored, gunmetal gray. Pentcheff held it in the cup of his palm and watched it squirm and flip in the California sun. On another day he might not have found it at all. Even though it was very small, Pentcheff understood immediately that the roly-poly was unusual. He'd found them here before, but this one was different. The bony plates along its curving back were covered with row upon row of small knobbly, peglike protrusions, visible to the naked eye. Since Pentcheff had never seen an isopod like it, he kept it.

"He brought the specimen in for Regina and me to look at," says Adam Wall, assistant collections manager for Crustacea at the Natural History Museum of Los Angeles County (NHM). Working with Regina Wetzer, associate curator and director of the Marine Biodiversity Center, Wall examined the specimen closely under high magnification and realized the extent of its uniqueness.

Described in 2015, the new species was named *Exosphaeroma pentcheffi*, after Pentcheff. Found in three feet of water on the Paseo del Mar shoreline, almost within sight of downtown Los Angeles, *E. pentcheffi* is a reminder that new species aren't found only in remote rainforests. It didn't come from the green, forested wall of an extinct volcano in Papua New Guinea. Instead, it lived undetected on the outskirts of one of the largest urban centers in the world.[1]

Almost nothing is known about its natural history. The extent of its range is not known either.[2] It might inhabit that one beach and nowhere else. In a similar discovery, entomologists at the Natural History Museum of Los Angeles County announced the discovery in 2015 of thirty new fly species captured in city backyards.[3] The project began as a bet between a museum trustee and entomologist Brian Brown, who claimed he could discover a new species in a Los Angeles backyard. Brown was a principal investigator for

BioScan—a citizen science initiative to study insect biodiversity. Thirty homeowners agreed to host backyard insect traps. The traps were set across the city—in Burbank and Glendale, in Hollywood, in Koreatown, and elsewhere. In the first weeks of the study, researchers began detecting unknown flies in the traps. There were thousands of flies. Entomologist Emily Hartop concentrated on flies belonging to a single genus: *Megaselia.*

Each of the traps yielded at least one new species—and each of them was named to honor the host of the site it came from. Thus *M. armstrongorum*; *M. bradyi*; *M. mikejohnsoni*; *M. steptoeae*; and so on.[4] Tens of thousands of other insect specimens were captured too—butterflies, moths, bees, mosquitoes, midges, flies, and gnats—but not assessed. In other words, unknown biodiversity is everywhere. We are surrounded by it. Go stand in your backyard and it's there.

Marine isopods are very common—ubiquitous, even. Before Pentcheff's discovery at Paseo del Mar, according to Wall, all marine isopods collected along the coast of California were routinely identified as *Exosphaeroma amplicauda.* And *E. amplicauda* had been known for a long time. It was described in 1857 by William Stimpson, an American naturalist. "He did the completely typical thing that you would have done back in 1857," says Wall. "He wrote literally one paragraph about it, he made one little drawing, and he made a couple of remarks about it. I think he says he found the specimen on a starfish, actually, and that it's relatively large, or something like that—very vague."

The description was so vague that any marine isopod that looked remotely like it was identified as *E. amplicauda.* But clearly there are other species too, like the specimen Pentcheff found. "It's pretty obvious that it's not *Exosphaeroma amplicauda*," says Wall. No one had really examined the isopods collected along the Pacific coast closely enough to identify novel species. "It's kind of a mess worldwide—this one little group. No one was willing to spend the time to distinguish between them. For many, many years taxonomists were calling anything that was in that genus the one species—*amplicauda.*"

Stimpson had been an industrious collector. Born in Boston in

1832, his primary interest was marine invertebrates. Almost none of the tens of thousands of specimens he collected in his lifetime still exist. By 1871 he was director of the Chicago Academy of Sciences. During his time there he built a vast natural history collection, one of the largest in the world. Then in October 1871 tragedy struck. The Great Chicago Fire burned for two days, killing three hundred people. When the fire reached the Academy of Sciences, the flames moved quickly through the supposedly fireproof museum buildings. Even the glass melted. Everything was destroyed: Stimpson's own specimens collected during twenty years of research were burned, as were ten thousand jars of crustacean specimens on loan from the Smithsonian Institution—at the time, the largest collection of its kind in the world. Countless written observations, half-completed manuscripts, and hand-drawn illustrations of specimens were all gone. Stimpson died a year later of tuberculosis.[5]

The class of Isopoda is large and diverse, including more than ten thousand species worldwide. About half of those are marine species, the rest are terrestrial. They're ancient animals, first appearing in the fossil record about three hundred million years ago, looking like miniature armored personnel carriers. Many isopods scavenge decaying animal and plant material; some are predators; others are internal or external parasites. The largest isopod of all, *Bathynomus giganteus*, is a pale pink lobster-sized deepwater species that lives on the seafloor and grows longer than two feet. The sphaeromatids alone—the family *E. pentcheffi* belongs to—include ninety-nine genera and almost seven hundred species.

When Stimpson identified *E. amplicauda*, he wasn't equipped with a modern scanning electron microscope. But Wall and his co-authors are. First, the roly-poly is dehydrated and coated with a mixture of gold and palladium. Then, by focusing a narrow beam of electrons on its surface, Wall can convert a tiny slate-colored specimen into a high-resolution image on a monitor screen. Researchers can explore its topographical contours, measuring its sculpted, down-curving tail fins and its seven pairs of jointed legs. Aided by Niel Bruce from the Museum of Tropical Queensland in Australia, Wall and Wetzer began borrowing specimens from other collections—isopods sampled from the length of the northeast Pacific coastline. "We were basically calling in museum specimens from anywhere with material that had been described as *Exosphaeroma amplicauda*," Wall says. "We ended up finding specimens from several different species."

Typically, Wall uses body ornamentation on the male isopod—like the rows of peglike protrusions Pentcheff found on the holotype from Paseo del Mar—to identify a new species. When Wall looked at the borrowed specimens he realized he was looking at several very distinctive species. Anyone who looks at the high-magnification images can tell them apart. Another new species, *E. russellhansoni*, comes from the shallow waters around Puget Sound, and *E. aphrodita* lives on the coastline off San Diego. Research has shown that *E. amplicauda*—the one species that suddenly became many—is probably restricted to the central California coast, near Tomales Bay.

From the National Museum of Natural History Collection in Washington, DC, Wall obtained another vial. Inside were twelve roly-poly specimens—dark gray, preserved in ethanol. The label on the vial read "Kyska Harbor, Alaska—Beach, Low Water." The specimens were collected in 1873 by William Healey Dall, an American malacologist and naturalist. Dall was aboard a primitive wooden research vessel among the remote Aleutian island chain in the Bering Sea. It was June, and the sea was dark. For months at sea, as part of a United States Coast Survey team, Dall took depth soundings and triangulated the Alaskan coastline, providing information that was used to make the first accurate nautical charts of the region. He also collected natural history specimens wherever he went. His field notes—handwritten in pencil in a slanted script in a little yellow logbook—have been digitized and are available online.[6] Reading them now, more than 140 years later, the notes have a strange sort of poetry, with entries like "Sitka. *Boschniakia glabra* dead in woods. Cabbages in bloom, abundant."

In the Rat Islands, an earthquake-prone group of volcanic islands toward the remote western end of the Aleutians, Dall collected the isopods. They came from Kiska harbor, an inlet on the east coast of

Kiska Island. A few weeks later, on August 5, 1873, he collected a few more, this time from the north coast of Amchitka, fifty miles to the southeast. On that day Dall's diary reads: "Aug 5, 1873: Rainy + stormy from SE with fog. . . . Attend to copying. In evening weather settles a little but wind remains the same."

In his field notes, Dall includes lists of the bird species he has seen or killed, checked off with a careless hand. He describes shooting a porpoise—from its range it was most likely a Dall's porpoise (*Phocoenoides dalli*), named after him. He carefully sketches shorelines and sloping cliffs. But he doesn't even mention the isopods. Eventually they were identified as *E. amplicauda* and stored away at the National Museum of Natural History. But they're not *E. amplicauda* either. Wall has named them *Exosphaeroma paydenae*. There are just a handful of them, he says—maybe a dozen. Most likely they are the only examples that exist in any collection anywhere. "*Exosphaeroma paydenae* has probably never been collected again," he says. "All the material for it came from that one collection. It's out there. It's on an island in the Aleutian Island chain—pretty remote."

The collection Wall oversees in Los Angeles has its own secrets waiting to be discovered. Herein lies the power of a well-tended collection: when a new technology is developed, the specimens can be investigated again in new ways.

"The collection at the Natural History Museum here in Los Angeles for Crustacea is the fourth largest in the world, and the second largest in the United States behind the Smithsonian," says Wall. "It's basically the taxonomic library of all the biodiversity we've managed to collect. There are several extinct species for which we have some of the very few representative samples."

Wall spends most of his time there in the monastic quiet of the collection, surrounded by jars of preserved specimens. He's certain the unidentified isopods are not alone. "I guarantee you there are hundreds if not thousands of yet-to-be recognized species essentially hiding in our collections," he says. "They're just chock-full of undescribed species."

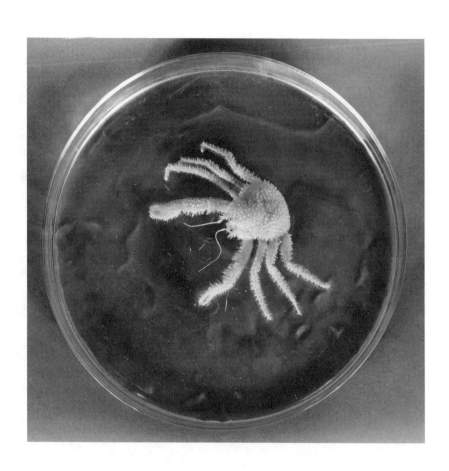

22

A Ball of Spines:
Makarov's King Crab
(*Paralomis makarovi*)

On June 4, 1906, the Bering Sea west of Alaska looked gray and hard like granite. A steamship called the USS *Albatross* was working a grid there, dredging the seafloor off Bowers Bank, a 430-mile submarine ridge that extends northward from the Aleutian Islands before curling westward toward Russia.

The *Albatross* was a research vessel owned by the United States Fish Commission. It was the first of its kind, built exclusively as an oceanographic research vessel and launched in 1882. In May 1906 it traveled north from San Francisco and docked at Dutch Harbor, Alaska. In the following months it would complete a circumnavigation of the northwestern Pacific Ocean: from Dutch Harbor westward across the Bering Sea to the Russian Far East, then south to Japan, threading a sinuous course between numerous Japanese islands before heading back to Alaska. Along the way the *Albatross* stopped at 385 stations to dredge the ocean floor for specimens, collecting hydrographic and meteorologic data at the same time.[1]

On June 4, 1906, at collecting station 4771, the plummet struck the ocean floor near Bowers Bank. It was 7:02 a.m. A half-mile of dark, cold water lay between the ship and the seafloor.

In 2008 Sally Snow (formerly Sally Hall) was partway through a five-week research visit to the Department of Invertebrate Zoology, at the National Museum of Natural History in Washington, DC, when she found a large fluid-filled jar. Inside were specimens from collecting station 4771. The contents were uncataloged: still unknown. Snow is a carcinologist—she studies crabs—and she was at the Department of Invertebrate Zoology to explore its extensive collection of king crab specimens. "I was doing an audit of their collections," she says, "looking at all the crabs within this particular family—Lithodidae."

For several weeks Snow, then a student at the National Oceanography Centre in Southampton, had been retrieving specimens of large decapod crabs from the collection and measuring them, building an array of precise measurement data for her dissertation. By comparing closely related crab species, she hoped to reveal how different species adapt to their habitats, particularly in relation to depth. "I was taking a specific set of measurements," she says, "particularly on the leg aspect ratios but also aspects of the carapace and how they were ornamented."

Snow had already crisscrossed the world visiting scientific collections to measure crab specimens: the Natural History Museum in London; the Muséum National d'Histoire Naturelle in Paris; the Museum Koenig in Bonn; the Naturmuseum Senckenberg in Frankfurt; the Alfred Wegener Institute for Polar and Marine Research in Bremerhaven, Germany; and other large invertebrate collections in Madrid and Vigo.

But now, at the Smithsonian Institution, she says, "They had large jars of unidentified species. Either they were identified to family level, so they had a label that said 'Lithodidae' and the collection details, or it just said 'Unidentified.'" The specimens from Bowers Bank, Snow says, belonged to a species she didn't recognize. She picked up a single specimen, its legs dangling limply from its carapace. At this time Snow had measured several thousand crab specimens. "I'd already been working on this research for about two and a half years, so I immediately knew it was a new species," she tells me. "It didn't look like anything I'd ever seen before."

By then, she estimates, she was one of only ten people in the world who could identify the specimens as examples of a novel species just by looking at them. Snow named the new species *Paralomis makarovi* after V. V. Makarov, who wrote a seminal 1938 paper on the biogeography of lithodid crabs. Morphologically, she says, the crab resembles a ball of spines. Every part of its compact body—its small, pear-shaped carapace and its legs, tucked neatly beneath it— bristles with spines. And each conical spine is covered with setae— a coat of stiff, hairlike fibers. The specimens are mostly pale pink, legs faded white like bone. "Based on other members of the family that live at that depth, I suspect they were dark pink or orange when alive," says Snow.

The crab belongs to a large family: the Lithodidae. About 107 species are known currently, in ten genera, and they are found in a range of habitats and ranges, mostly pitch-black, abyssal, deepwater environments. Technically lithodids are not considered true crabs, Brachyura; they're more closely related to hermit crabs.

The lithodids remain mysterious. The great depths where they live make them difficult to study. Another species, *Paralomis bouvieri*, has been found at depths of 4,152 meters, more than two and a half miles deep, where it's almost impossible to study its natural history. In fact, *P. makarovi* doesn't look much like a typical lithodid at all. First, it's small. Many king crab species have earned their names: they grow to enormous sizes. *Paralithodes camtschaticus*—the red king crab—is a heavily harvested species also found at Bowers Bank, sharing its range with *P. makarovi* and several other lithodid species. Its broad rust-colored carapace is ornamented with rough nodules, covered and ridged with tubercules and knobby protrusions. The carapace of a large old *P. camtschaticus* specimen can be almost a foot across—the size of a hubcap. It has a six-foot leg span, more typical of lithodids. "I was used to looking at massive crabs," says Snow.

In contrast, *P. makarovi* measures about ten centimeters across at its widest point. It fits comfortably in Snow's palm like a pale, bristly disk. When she first saw the specimens in the jar, she suspected they were juveniles of an already known lithodid species. As they develop into adults, crabs change a lot in appearance, Snow says, so she asked researchers she knew in Alaska if they had seen crabs like the specimens in the collection. They had, they told her, but the species had not been described. It had no name. "King crabs have been known for a long time. To find something that was so different from anything already known was very exciting."

In her paper published in *Zootaxa* in 2009, Snow described three more novel species too: *Paralomis alcockiana*, from South Carolina; *Paralomis nivosa*, from the Philippines; and *Lithodes galapagensis*, from the Galápagos archipelago.[2] She found the holotypes of them all at the National Museum of Natural History. *P. nivosa* had spent almost as long on the shelf as *P. marakovi*. It was collected in December 1908, near Palawan in the Philippines, during a subsequent voyage of the *Albatross*.

Discoveries like Snow's take place all the time in established scientific collections like the Smithsonian Institution. Put simply, the invertebrate collection at the National Museum of Natural History contains so many specimens that it is just impossible to catalog them all.

"The Crustacea collections in the National Museum of Natural History contain about 631,779 lots, meaning jars or containers of various sizes," says Rafael Lemaitre, carcinologist and curator of Crustacea at the museum. About 260,000 of these lots, he says, or fewer than half, are identified to some taxonomic level. The rest are unidentified. Twenty years ago, the National Museum of Natural History employed ten curators for its collection of crustaceans, but now Lemaitre is the only one. Like a lighthouse keeper, he alone walks the quiet aisles and peers into the jars, carefully tending to the specimens—more than a half-million lots, most of them awaiting study.

In 1906 the USS *Albatross* represented a paradigm shift. It industrialized scientific collection. There was a state-of-the-art laboratory onboard to process the captured samples the moment they surfaced. The expedition was bookended by disaster: first its departure was delayed for several days by the San Francisco earthquake that May; then on November 21, 1906, during the return leg, Lieutenant Commander LeRoy M. Garrett, the ship's captain, was washed overboard northwest of Honolulu and drowned.[3] Voyages were perilous but very productive. Scientists worked around the clock processing samples. There was even an ornithologist on board to collect and catalog the bird specimens. Austin Hobart Clark kept meticulous notes, writing about Canada geese on Agattu, in the Aleutian Islands: "I shot nine, using merely a very light charge of powder with an ounce of No. 10 shot which I had brought with me for the purpose of securing specimens of song sparrows (*Melospiza*) and longspurs (*Calcarius*); even with that light load I secured two at one discharge of my 12-bore. I believe I could have killed a hundred or more in the course of a morning's walk had there been any object in so doing."[4]

When the *Albatross* surveyed the Philippines in 1907, naturalists aboard used bottom trawls, dredges, midwater trawls, gill nets, plankton nets, beach seines and hand lines. They threw dynamite into the water to collect reef fish, blowing apart most of the specimens. Occasionally they even poisoned tide pools with copper sulfate to collect the organisms found in them. All together, between 100,000 and 150,000 specimens were collected—about 8 percent of the entire collection of the National Museum of Natural History's Division of Fishes. Several species were collected for the first and last time during the 1907 expedition—like the congrid eels *Congrhynchus talabonoides* and *Bathyuroconger parvibranchialis*.[5]

When biological material from the 1906 expedition to the northwestern Pacific began to arrive at the National Museum of Natural History—an estimated ninety cases of it—there were too many specimens to evaluate and too few taxonomists to do the work. This barrier to identification has only worsened in the past decade. Lemaitre cannot describe and name the thousands of unknown specimens by himself. His situation is typical. Many large institutions have lost curators. Collections at smaller, underfunded institutions are sometimes lost altogether—basically orphaned. Occasionally, orphaned collections are subsumed into larger collections.

Sometimes they deteriorate and are simply forgotten. Specimens are discarded. Labels are lost. When that happens, decades and sometimes centuries of effort disappear too. Before the Enlightenment, says taxonomist Quentin Wheeler, our knowledge of species doubled every four hundred years. By the time Linnaeus published the tenth edition of *Systema Naturae* in 1758 and introduced the concept of binomial nomenclature, taxonomic knowledge doubled much more quickly—about every fifty years. It was the golden age of species discovery. But by the mid-twentieth century the rate of naming novel species had slowed again. Currently it doubles about every two hundred years. In other words, our rate of discovery has reverted to what it was in the 1750s.[6] This is not because there are fewer species left to describe: we have barely begun to describe and classify life on Earth. By conservative estimates, taxonomists have named only about 10 percent of the planet's biodiversity. Unknown species wait. We need to fill in the gaps by describing species.

In all, more than two hundred novel species were described from specimens collected during the 1906 *Albatross* expedition to the northwest Pacific Ocean. Also dredged from station 4771 were *Lithodes couesi*, an enormous scarlet plate-sized king crab; numerous specimens of isopods, shrimp, hydrozoans, and octocorals; an ophiuriod, or brittle star; a single specimen of *Stauroteuthis albatrossi*—a small, flattened species of benthic octopus; a dried *Henricia leviuscula spiculifera*, or blood star; six squat lobsters; more than sixty *Eualus biunguis* shrimp; a novel, undescribed species of sponge; and two specimens of the salmon glass-scallop, *Cyclopecten davidsoni*, a benthic bivalve. It's not surprising that a novel species of lithodid sat undetected for a century—stored away in an undisturbed corner of the collections. Sally Snow can't discern one species of isopod from another—she says she doesn't even know what they are. But she noticed *P. makarovi* immediately, her eyes keen to its uniqueness.

"The unworked collections are like a treasure," says Lemaitre. "Every time we look at those specimens it's like exploring the unknown."

Botanical

23

In an Ikea Bag:
The Custard Apple Family
(*Monanthotaxis* Genus)

In April 2015 French botanist Thomas Couvreur, a researcher at
the Institute of Research for Development in Montpelier, France,
was walking in the Ottotomo Forest Reserve in central Cameroon.
He was near a main road, in a botanically well-known area he'd surveyed many times before. He didn't expect to find any new plant
species in such a well-surveyed part of the rainforest. He hadn't
even brought his plant press—an essential part of the kit botanists
take with them into the field, which allows them to preserve specimens permanently.

A short distance from the path, Couvreur stumbled over a liana—a long woody vine with a thin reddish trunk. It was covered
with delicate, small-petaled ivory flowers. Couvreur, a systematist who studies the relations between different plant species, had
been living in Cameroon for almost four years. By sight, he knew
the liana belonged to the *Monanthotaxis* genus. He was less certain
about its species. "Finding good flowering material of this genus is
quite rare," he says, "so I collected it in an Ikea bag." Later, back in
Yaoundé, Couvreur removed the liana from the blue plastic bag and
carefully placed the specimen in the plant press.

In 2016 Dutch researcher Paul Hoekstra described the plant, naming it *Monanthotaxis couvreurii* to honor Couvreur for collecting the holotype. A researcher at Naturalis Biodiversity Center and Wageningen University in the Netherlands, Hoekstra described eight more *Monanthotaxis* species in the same paper.[1]

The *Monanthotaxis* genus belongs to the Annonaceae family—otherwise known as the custard apple family. Its members are found in the tropics, where they mostly grow as lianas. Their woody vines extend vertically into the rainforest, insinuating themselves into a tangled network among the branches that form the canopy. High in the air—in some instances the trees they grow in can be more than 120 feet tall—the flowers bloom and quickly disappear. A botanist like Couvreur, working far below on the forest floor, might never see them. "They're just up there and you can't collect them, so the whole genus is relatively undercollected," he says. "We have very little material."

This presents a particular problem for botanists. Imagine if the species-defining characteristics of a mammal like the olinguito lasted only a few days, then disappeared. When a flowering specimen is found in the field, a botanist will scramble to collect it in time. A couple of years earlier, in November 2013, Couvreur had collected another *Monanthotaxis* liana—this time in Gabon, near the Ipassa Research Station in Ivindo National Park. "This liana had actually fallen to the ground," says Couvreur. "It was hanging over the trail. I could see that it was *Monanthotaxis*. In this case the flowers were not on the trunk, they were at the top of the branches. As the liana fell down, I had easy access and could collect them."

Some species—*M. couvreurii* is an example—grow flowers on their trunks. They are known as cauliflory species. They present another very specific challenge to botanists: the flowers might grow on the trunk, but the leaves, with their own important distinguishing characteristics, are far away in the canopy, at the end of the liana. Other species, like the plant Couvreur saw in 2013 in Gabon, belong to a larger group of species in the *Monanthotaxis* genus that grow flowers at the ends of their branches, at the distal end of the liana. Studded with budding flowers like little green peas, the fallen liana later became the holotype for *Monanthotaxis latistamina*—another of the new species Hoekstra named.

According to a 2010 study, only 16 percent of new plant species are described within five years of their collection in the field. The rest wait much longer. Almost a quarter of new species in collections wait more than fifty years for description. They are dried and pressed flat on herbarium sheets—a method mostly unchanged from the techniques used to preserve plant specimens in the 1500s and even earlier. They are given catalog numbers and accessioned into collections. The oldest botanical specimens in the United States National Herbarium at the Smithsonian Institution were collected in 1504.

Old botanical specimens are beguiling. They combine the precision of the scientific method with the organic matter-of-fact simplicity of the physical object itself. In the past, herbarium sheets were often bound together in book form. The pages of four-hundred-year-old notebooks are covered with the faded green-gray scrollwork of flattened ancient leaves, which unfurl wildly toward the corners.

Among the botanical specimens at the Naturalis Biodiversity Center in Leiden is the first known example of a tomato plant, collected in the early 1540s. It includes a wrinkled, misshapen tomato, its vibrant colors preserved, dangling forever from its stalk. Written beneath the plant is the name *Poma amoris*, or love apple—an early name used for the tomato as it spread across Europe. It is a true antiquity. For context: when the tomato was pressed to the page, Shakespeare hadn't been born. About the same time Henry VIII, in his fantastic rotundity, was busy killing his wives. He executed Catherine Howard in 1542 and married his sixth and final wife, Katherine Parr, a year later.[2]

While examining more than two thousand *Monanthotaxis* specimens borrowed from herbaria around the world for his revision, Hoekstra found archived specimens of all nine of the new species he has described. Couvreur had collected the holotype for *M. couvreurii* in Cameroon in 2015, but older specimens had been collected in May

1970 and deposited in the Muséum National d'Histoire Naturelle in Paris and at the National Herbarium of Cameroon. Specimens of *M. latistamina* were in the Paris collection too, dating back to 1975. And so on. Most of the other specimens were collected in the 1960s and had been part of herbaria for fifty years. Earlier generations of taxonomists had struggled to discriminate between different *Monanthotaxis* species because archived specimens lacked the species-defining flowers or fruit: they were sterile. Instead, botanists had to study the leaves, which generally refused to reveal their uniqueness. With the advent of molecular techniques, says Hoekstra, that too has changed.

"I look at morphology," he says, "but I also compare specimens at the DNA level." By comparing several DNA markers, Hoekstra can understand how the *Monanthotaxis* species are related to one another. To do this he needs only a leaf, or even just a fragment of a leaf—usually not informative enough to delimit a species based on morphology alone.

Among the newly named species, *Monanthotaxis zenkeris* spent the longest on the shelf. The holotype was collected in 1907 in Cameroon by its namesake, Georg August Zenker, a German gardener and botanist. On recent field expeditions to Cameroon, Hoekstra hasn't seen the plant. He suspects it has disappeared in the wild; all we have now is the old archived material. When he collected the holotype in 1907, Zenker had been living in Africa for more than twenty years—first in the Congo, then in Gabon. Finally, when he settled in Cameroon, a German protectorate, he married an African woman and learned to speak the local language. Zenker was taught to understand the drummed messages that pulsed through the rainforest. In time he became one of the people he went to live among. But he was a collector too. Several species have been named after him. He collected all things, avidly. After he visited Zenker in 1897, Jesko von Puttkamer, the governor of German Cameroon, wrote: "His house was a complete museum full of ethnographic curiosities, photographs, oil and watercolour paintings, herbaria, furs and skull, weapons, fetishes, stuffed birds, etc."

The iDigBio online database of biological material includes more than four thousand specimens collected by Zenker. They all came from the same small corner of Cameroon. Now they are in collections in Oslo, Paris, and São Paulo, and in the Field Museum, the New York Botanical Garden Herbarium, and the Museum of Evolution in Uppsala, Sweden, which houses specimens collected by Linnaeus. There is a photograph of Zenker in Bipindi, in southwestern Cameroon: tall and slim, dark-eyed and black-bearded, he's wearing a white peaked cap and elegant white high-waisted trousers with suspenders, the long fingers of one hand draped languidly over a leg. Who would dare to wear all white in the Cameroon rainforest?

This is exactly how natural history collections are supposed to work. The specimens in herbaria represent a conversation that takes place over centuries. A specimen is meaningless by itself, assessed out of context. The number of *Monanthotaxis* species in 2014 was fifty-six; now it's sixty-seven. Soon there will be even more, says Hoekstra. He suspects the number is closer to a hundred, and he plans to describe several more species over the next few years. As he does, systematists like Couvreur will begin to gain a more complete and nuanced understanding of the interrelations between the species they study, and the forces that drove their evolution—how these species came to be.

This is important. Currently botanists understand a fraction of what goes on in vast interconnected ecosystems like the Cameroon and Gabon rainforests. The rainforest is like a sprawling green brain. But we understand it less than we understand an actual brain. Researchers like Hoekstra still don't know some of the most basic facts about *Monanthotaxis* species, like how they pollinate, how their seeds are dispersed into the environment, when the different species diverged genetically, and how they radiated across Africa, evolving as they went. Botanists have never even seen or collected the fruit of several of the new species. Other species may contain pharmacologic agents that could be useful in treating malaria and other diseases. Hoekstra tells me it's easier to list what we know about the genus that what we don't.[3]

The oldest known *Monanthotaxis* specimen was collected in

Sierra Leone in December 1795 by Adam Afzelius, a Swedish botanist and apostle of Linnaeus. Hoekstra still has the specimen on loan
in Leiden. It came from the herbarium at Uppsala, in Sweden. At
more than 220 years old, the specimen is brittle and delicate now,
but still useful. According to Hoekstra, it belongs to *Monanthotaxis
barteri*, but it was misidentified for centuries. For a long time the
key to identifying West African *Monanthotaxis* species was their
floral characteristics, says Hoekstra. When Afzelius collected the
specimen, it was fruiting instead. A few weeks earlier it would have
been flowering and recognizable, but when he saw it in the forest
it was impossible to identify. Hoekstra can identify it now, though.
Sometimes the classification of a species takes centuries, requiring
one small refinement after another.

As for Couvreur, he continues to spend as much time as possible
in the field. In 2015 he described a new plant species from Gabon—
a pretty whorl of little red petals around a yellow cone-shaped
center of loosely arranged stamens. It's very distinctive, and Couvreur found it near a main road in another well-surveyed region.
He named it *Sirdavidia solanonna* to honor Sir David Attenborough.
But later, at the herbarium at Kew Gardens in London, he found a
dried specimen of the plant collected in 1973.

His work in the field—months beneath the canopy in rainforests
across Central Africa—only provides the raw material. The specimens are questions. The collections provide the answers. "They're
the basis of everything," says Couvreur. "Without them, we wouldn't
know anything about anything."

The Others

Waiting with Their Jackets On: The Fossils (Paleontology Specimens Collected by Elmer Riggs)

Bill Simpson stands at the controls of a forklift in a storage room dedicated to the large fossils at the Field Museum of Natural History in Chicago. On the wooden pallet in front of him are two long slate-gray bones: *Triceratops* femurs, laid alongside each other like felled tree trunks. They were collected in Montana by paleontologist Elmer Riggs during a 1904 field expedition. Simpson, who has worked at the Field Museum since 1979, estimates the bones each weigh about a hundred pounds.

Out in the field, when paleontologists find fossilized remains they wrap them with bandages or burlap soaked in wet plaster, which dries to form a protective cover around the delicate fossil. Elmer Riggs did it to the *Triceratops* femurs in 1904, and modern paleontologists still do it. But when Riggs unearthed the bones in Montana, it was a particularly important step. Taken from the dig site in the bed of a horse-drawn wooden wagon, the bones were jostled and bounced across the uneven ground. There is a photo of Riggs on expedition, riding a small white horse in front of a wagon drawn by two horses across a landscape littered with boulders. He holds aloft a flag emblazoned with the words Field Columbian Museum.

Many decades ago, the *Triceratops* bones arrived at the museum encased in their field jackets with their field numbers written on top. Fossil preparators then carefully freed the bone from the plaster, removing the surrounding matrix from the fossil rock by rock. "The process of taking rocks off the fossil is a real bottleneck in our science," says Simpson. "We don't prepare every fossil specimen we collect. We don't have the staffing for that."

Instead, low priority field jackets wait. To Simpson's left, lining one wall, are several rows of enormous field jackets. They're arranged on shelves in the cavernous room Simpson calls the oversize range. Riggs collected the contents of some of the jackets more than a century ago—in the Rocky Mountains, in Alberta, Canada, and while on expedition in Bolivia and Argentina. "When he came back from expedition only a certain number of the things he collected were prepared, including big things that were in field jackets," Simpson says. No one knows for certain what's in the others.

In a photo taken in December 1899, a well-dressed Riggs stands in the paleontology laboratory at the Field Columbian Museum. His assistant Harold W. Menke, wearing a striped apron, stands across the table from him, removing the field jacket from a fossil collected at Grand Junction, Colorado. Behind the two men an enormous black femur stands propped cartoonlike against the wall. The end of the femur is as large and rounded as Menke's head. All around the two men are fossils: oversize vertebrae, long bones, a scapula as big as an oar. On one side of the room, still in its plaster field jacket, stands a *Brachiosaurus* femur. If a *Brachiosaurus* femur is already enormous, clad in a white, bone-shaped plaster jacket it looks outrageously so—like the bone Fred Flintstone throws on top of his car in the opening credits of *The Flintstones*.

A mammal fossil specialist, Riggs was hired at the end of 1898 by the Field Museum—then still named the Field Columbian Museum. He began working there at the end of a very specific period in paleontology known as the Bone Wars. For the previous two decades, two paleontologists—Edward Drinker Cope of the Academy of Natural Sciences in Philadelphia and Othniel Charles Marsh from the Peabody Museum of Natural History at Yale—had waged an unfriendly and mean-spirited contest to become the world's preeminent fossil collector. The battle was fought across the American West, through the Rockies, in quarries and rock formations across Wyoming, Utah, Colorado, and elsewhere. The results: *Allosaurus* (Marsh), *Apatosaurus* (Marsh), *Dimetrodon* (Cope), *Camarasaurus* (Cope), *Coelophysis* (Cope), *Stegosaurus* (Marsh), *Triceratops* (Marsh), and others.[1]

Between them Cope and Marsh—rivals to the end—discovered more than thirty dinosaur species. Before Riggs was hired in 1898, he was asked to prove his worth: "They took him out in the field to the Badlands of South Dakota to give him a trial run," says Simpson. He succeeded, excavating numerous fossils. "He was our first paleontologist. In 1899 he went to Wyoming; in 1900 and 1901 he collected in western Colorado. That's where he got *Brachiosaurus*."

Riggs had been set a specific task: excavate a complete dinosaur skeleton to put on display at the Field Museum. At the turn of the twentieth century, four museums in the United States were locked in another battle. Each museum wanted to be the first to display a complete mounted skeleton of a sauropod dinosaur. The race had begun. "It consisted of the American Museum in New York, the Yale Peabody in New Haven, and the Carnegie Museum in Pittsburgh," says Simpson. "The Field Museum was a late, less well funded entry."

Riggs went west in search of long bones. "In 1901 he dug up about 65 percent of a *Brontosaurus*, which they then started taking the rock off and bending the armature for, to put it on display." In 1905 the American Museum and Yale Peabody put their *Brontosaurus* specimens on display. Eventually the *Brontosaurus* became better known as *Apatosaurus*. The dispute over whether *Brontosaurus* is a junior synonym of *Apatosaurus* or a valid species in its own right continues. The Carnegie Museum followed with a *Diplodocus*, a close relative of *Apatosaurus*, in 1907. Finally, a year later the Field Museum put its own *Apatosaurus* on display. In 1922 Riggs led an expedition to southern Alberta, Canada, to collect late Cretaceous dinosaurs from the remote Red Deer River badlands. In all, Riggs and his five-man team excavated numerous fossils, including prehistoric turtles and several duck-billed dinosaurs. One of them, a *Lambeosaurus*, is part of an exhibit at the Field Museum. Its bones are laid out in situ on the ground. The skeleton of a *Daspletosaurus* towers above it, posed as if about to eat it. But other fossils collected in 1922 remain unprepared, says Simpson. "There are still quite a few field jackets containing mostly duck-billed dinosaurs." They sit on shelves in the oversize range room like strange white boulders. "Duck-billed dinosaurs have been well studied and are fairly low priority.... Even though we do have a dinosaur paleontologist now, he doesn't focus on duck-billed dinosaurs, so those specimens that were collected in 1922 are still in these big field jackets."

In 1999 Simpson was in charge of preparing Sue, the *Tyrannosaurus rex* skeleton that now dominates Stanley Field Hall at the Field Museum. "I had two preparation laboratories," he says. "They were both what we call exhibit labs, where they have a wall of windows so the public can see what we're doing." One of the labs was at the Field Museum, the other was at Disney World in Orlando, Florida. "I sent the back half of Sue down to Disney World to be prepared by our three preparators down there." Several months ahead of schedule, the Florida team finished preparing the bones they'd been sent. "I wanted to make sure we were getting full use of their expertise for the final two or three months of their contracts. I looked through Riggs's field notes for possible gems that were still in field jackets."

One of them looked promising: "I found reference to a small carnivorous dinosaur that he'd collected in 1922," says Simpson. It had come from the expedition to the Red Deer Badlands. "This would be either a juvenile of a *Tyrannosaurus* or some small adult, maybe some new species, of carnivorous dinosaur. Either way it would be interesting. There were three field jackets. I sent those three field jackets down to our preparators at Disney World."

Inside the three plaster jackets, the preparators found part of a *Gorgosaurus* skeleton. "This little juvenile *Gorgosaurus*—its tail was even more complete than Sue's. Its two hind legs were even more complete too," Simpson says. "One of the hind legs was 100 percent complete. Every single bone, including all the little bones in the foot, was there. It was really an amazing little specimen."

It's not only bones hidden in field jackets that go unnoticed. In October 2016, Field Museum vertebrate paleontologist Susumu Tomiya named two new species of extinct beardogs—small-bodied carnivorous relatives of dogs and bears that lived an estimated forty million years ago. Tomiya named one of them from a jawbone and an incomplete fragment of skull in the Field collection.[2] The fossil hadn't been excavated by Riggs; it was discovered in southwestern Texas in 1946. It wasn't still in its field jacket—Tomiya found it in a drawer. The specimen had been incorrectly classified in the 1980s and placed in a genus it didn't belong to.

Tomiya found the jawbone by accident. "In my spare time I like to walk around the aisles in the collections and open up drawers," he told the Associated Press. "One day I just stumbled on these interesting-looking jaws of a little carnivore." The find is an important one. Scientists have now described about half a dozen species that belong to the amphicyonids—the beardogs. The new data will help them unravel how the group evolved.

In 2003 the vertebrate paleontology collection was transferred to a new wing of the Field Museum—a cavernous 180,000-square-foot, two-story underground facility. By comparison, the space previously used to house the unprepared specimens had been cramped and poorly lit. "It wasn't big enough," says Simpson. "The field jackets were piled up four deep. There wasn't enough light in the room, but even if there had been you couldn't have read the field numbers on most of the jackets because they were buried by the ones on top. When we moved into this new facility, which is state of the art, for the first time I was able to do a complete inventory of our field jackets. We discovered things we didn't even know we had."

In a 2014 paper in *Historical Biology*, Field Museum paleontologists Michael Hanson and Peter Makovicky described the discovery of the fossilized foot bones of a *Torvosaurus* from one of the field jackets: specimen FMNH PR 3060.[3] The bones came from the Freezeout Hills in Wyoming, collected by Riggs in the summer of

1899. About the size and shape of a microwave oven, Simpson says, the field jacket had survived more than a century and was damaged, losing much of its outer plaster. "These old worn-out field jackets have all the stiffness of a dirty sock, and the fossils inside are no longer protected," he says.

Rather than patch and repair the jacket—during the move Simpson used more than half a ton of plaster to restore other damaged field jackets—he decided to schedule its preparation instead. *Torvosaurus*, a large carnivorous dinosaur, was first described in 1979 from bones discovered in a quarry in northwestern Colorado in 1971.[4] The bones Riggs collected in 1899 had been found much earlier but had remained unexamined.

In another instance, Simpson found thirteen huge mystery field jackets—large blocks of plaster for which he had no data. "A couple of them did have numbers," he says, "but I just couldn't interpret them." Laying the field jackets out on the ground, he attempted to decipher their contents. On the outside of one of the field jackets, someone had scrawled a number: F-742. "At one point I recognized that five of them seemed similar to one another—like their cross sections," he says. "I arranged them together on the floor and realized we had a six-foot-long femur from a big—what used to be called—*Brontosaurus* that was in five field jackets. That helped me interpret the remaining specimens—they were big back bones from the same specimen. In fact, one of them was D-742, which meant dorsal, the dorsal vertebra, and the F-742 meant femur. So 742, I then realized, was the date: July 1942."

These bones, too, remain encased in their protective jackets—future spare parts for the *Apatosaurus* skeleton that has been on display since 1908, says Simpson.

Elmer Riggs died in 1963 at ninety-four. His career is now defined by two South American expeditions—one in 1923 and another in 1926. He collected over 1,500 fossils from Bolivia and Argentina, including over 1,300 fossil mammals, 126 fossil birds, thirteen fossil reptiles, including a few dinosaur bones, and several fossil fishes.

The South American collection, particularly the fossil mammals, is studied every year by paleontologists from all over the world. Among the specimens are nineteen holotypes of newly described species: several prehistoric rodents, giant ground sloths, carnivorous marsupials, and two large species of flightless predatory birds called terror birds. The American public followed Riggs as he went via reports in newspapers.[5] A February 1925 article in the *New York Times* carries the headline "Field Museum Explorers Find Battleground of Monsters in Bolivia."

In total Riggs returned with more than three thousand pounds of fossils. At one time or another they all arrived at the Field Museum encased in their coatings of burlap and plaster—unpolished gems wrapped in stiffened dirty socks. A mammal specialist, Riggs spent the rest of his career collecting fossil mammals. In 1934 he described *Thylacosmilus atrox*—literally, "terrifying pouched carving knife"—from fossil remains he collected in 1926 in Argentina. A large saber-toothed carnivorous marsupial, *T. atrox* was a Pliocene mammal. When Riggs excavated its broad, heavy skull at Catamarca in northwestern Argentina, it had been extinct for an estimated two million years. The skull is dominated—eclipsed, almost—by two enormous sickle-shaped canine teeth that sit in grooved channels in its strangely adapted lower jaw. Almost a century later, paleontologists continue to make discoveries from the enormous amount of material Riggs collected in South America. In 2012 American Museum paleontologist Bruce Shockey described *Elmerriggsia fieldia*, a new species of hoofed mammal called a toxodont, from fossils Riggs excavated in 1924 at Pico Truncado, Argentina.[6]

Standing at the controls of the forklift, Simpson carefully inches forward, replacing the pallet with the triceratops femurs arranged on it, slotting it into its designated place on the shelf. On the other side of the room an array of unopened field jackets awaits: *Apatosaurus* femurs, duck-billed dinosaurs, perhaps another, even more complete *Gorgosaurus* skeleton with every bone, even the smallest, intact. In total, says Simpson, from Riggs's South American expeditions alone there are twenty-eight plaster field jackets and thirty-five other packages that remain unopened at the Field Museum.

Arranged in chronological order on the shelves, they are in storage deep beneath the busy streets of Chicago. They have returned belowground once again, into the quiet.

"We have lots of field jackets of mammals, big mammals, that he collected," Simpson says. "There's a chance that in one of those jackets there might be something new." Until someone finally cracks them open and begins to pull away the burlap, they wait on their pallets, stacked like pale ghosts.

25

The First Art: The Earliest Hominin Engraving (a 500,000-Year-Old Shell)

In the collection at the Naturalis Biodiversity Center in Leiden, in the Netherlands, there is a fossilized shell. Pale gray and curved like a shallow bowl, it belongs to the now-extinct species *Pseudodon vondembuschianus trinilensis.* In total there were hundreds of shells in the collection, all piled in boxes, says Jose Joordens, an archaeologist at the University of Leiden. "This freshwater shell was one of many—more than a hundred all together." But this shell was different from the rest.

In 1891, at Trinil on the Indonesian island of Java, Dutch army physician Eugène Dubois was excavating a site near the Solo River. He had come to Asia for one reason: to find the remains of the missing link—the transitional species that bridged the gap in the fossil record between apes and modern humans. Dubois had been inspired by the discovery of two almost complete Neanderthal skeletons in Spy, Belgium, in 1886. About this time, anthropologists like Ernst Haeckel had settled on the theory that human life originated in Asia, not Africa. After joining the Dutch army, Dubois arranged to be stationed in the Dutch East Indies, in present-day Indonesia, where he could search for the fossilized remains of transitional hominin species.

People had found fossils at Trinil before, and Dubois decided that was where he would begin to dig. First, near a wide bend in the river he found an incomplete skull—just the calotte, or cranial cap. It had a thick, bony brow ridge; the rest of the skull was gone. A few feet away he uncovered two large molars. A year later, still farther away, was a single unbroken femur—long and humanlike. DuBois had found the missing link. It became known around the world as Java Man. It had a smaller brain than ours—about two-thirds the volume of a modern human's—but it walked upright like us. Dubois named it *Pithecanthropus erectus.* The find was announced in 1894. Later, as the taxonomy of early hominins was reorganized, the species was renamed *Homo erectus.*[1]

Along with the bones and teeth, Dubois collected hundreds of fossilized shells found at the dig site at Trinil. There are 143 articulated shells and other fragments in the Dubois collection at Leiden. Together they represent eleven species of mollusks. On the exterior of one of the shells, across its rounded belly, are several deep, angular scratches. "It's more or less a zigzag pattern," says Joordens. "Each angle is very much like the other. It's a very regular pattern. It sits more or less in the middle of the outside of the shell."

The marks were made about 500,000 years ago by *Homo erectus*, the only inhabitant of the island at that time, Joordens says. Geometric engravings like the grooves on the shell are indicators of modern cognition—a creative and expressive act. Purposeful, even art, perhaps—and at the very least proof of a large and complicated brain acting with intent. The scratches on the shell were discovered by accident during a 2007 visit by Stephen Munro, a National Museum of Australia archaeologist who studies shells excavated from early hominin sites. He almost missed them.

Munro carefully removed the shells from the cigar boxes they had been stored in for more than a century. Since he was in Leiden for only a few hours and wouldn't have time to inspect every shell, he decided to photograph them all instead. He quickly placed shells beneath a camera and photographed hundreds of them. Then he left. More than a month later, when Munro returned to Australia from fieldwork, he still hadn't looked at the photographs. All together, he says, there were about three hundred digital images. He started scrolling through them, inspecting the shells for signs of human agency in the fracture patterns on their surfaces. Then he opened one of the last images: number 298.

"Nothing prepared me for the pattern of cut marks I saw when I opened image 298," says Munro. "It was so startling I almost fell off my chair." Munro leaned closer to his monitor, nose almost touching the screen, to inspect the image of specimen DUB1006-fL—a *Pseudodon vondembuschianus trinilensis*. The markings on it, two canted lines next to a strange slanted *M*—like this: *I I M*, scratched across the surface of the shell. "Shells this old simply weren't supposed to have marking like this, and *Homo erectus* wasn't supposed to be capable of creating them," he says. The markings are so faint that Munro hadn't seen them when he was handling the shell a month before. Back then the shells had been lit from the side with oblique light, which had made the markings invisible. "I knew immediately that, should they turn out to be genuine, this would be a hugely significant discovery," he says.[2]

The shell represents the earliest abstract markings made by any hominin species, and by a very wide margin. It gets at the root of who were are. Paleolithic cave paintings like those at Lascaux in France and Altamira in Spain date to about forty thousand years ago. Recently discovered Neanderthal markings—a hashtag-shaped carving on a cave wall in Gibraltar—have been dated at about thirty-nine thousand years ago. The marking on the shell is far older. Found in a wide arc from Georgia in Eastern Europe eastward as far as Java, *H. erectus* was an enigmatic species of hominin. By the time Neanderthals and early *H. sapiens* were painting on cave walls in Europe, *Homo erectus* was already extinct, disappearing an estimated 100,000 years earlier.

"*Homo erectus* probably originated in Africa about 1.9 million years ago," Joordens says. "By 200,000 years later it was more or less spreading out of Africa. It was the first species that really looked a lot like us: it had long legs; it had a similar build and body shape; it had quite a large brain, so in many respects you could say it is the first species that is a transition between our australopithecine ancestors and what looks more like a modern human."

Using two dating methods, Joordens and her co-investigators dated the sediment contained within the cavities of the shells and found it was between 380,000 and 640,000 years old. She suspects *H. erectus* used a readily available sharp object—a shark tooth, perhaps—to make the markings on the shell.[3] Several other shells in the collection bore a small rounded hole near the rear adductor muscle of the mussel, says Joordens, left behind when *H. erectus* opened the shells to eat the contents. Another of the shells had been used as a tool. "The sides were modified a bit so that it had a very sharp cutting edge. There are still a few more shells in the collection I'm currently studying that may also be tools. It shows that once you start looking you see many more things."

The shell might be one of the most important objects we have as a species. It vibrates with meaning. It precedes us by such a long time that its existence represents the widest sweep of history—a sweep so wide that most of it is actually prehistory. A mark on a shell, made by hands like ours, is like a faint glimmer in the darkness, a flash of lightning out at sea. When it was made half a million years ago, the ancestors of Neanderthals—*Homo heidelbergensis* or *Homo rhodesiensis*—were beginning to leave Africa, radiating outward vast distances across the world. The Neanderthals survived until an estimated thirty thousand years ago.

In 2003 archaeologists uncovered the skeletal remains of another small-bodied hominin species on the island of Flores, seven hundred miles east of Java. Named *Homo floresiensis* and nicknamed hobbit, the bones, and the tools found near them, are dated to between about 100,000 and 50,000 years ago.[4] In Siberia another hominin species, known collectively as the Denisovans, existed until at least 50,000 years ago.[5] There were other early hominins scattered around too—perhaps even a dozen or more species roaming the landscape—but they all are gone now. Perhaps in another version of events *H. erectus* might have risen to prominence instead of us. The survival of *H. sapiens* was never assured. When *H. erectus* held a pseudodon shell in its hands on Java and scored its surface with a shark tooth—fresh white lines left on a moss green shell— the entire world was still in play: *I I M.*

Joordens has returned to Trinil for more fieldwork: digging alongside flooded rice paddies fringed with palms. The peak of a volcano, Mount Lawu, rises green and hazy in the distance, inactive since 1885. Who knows what she might find there on the wet riverbanks? Joordens is hoping to find more artifacts—more marks of human agency. She hopes their importance will be noticed immediately. If not, collections like the ever expanding archive at the Naturalis Biodiversity Center will retain them. The specimens will be maintained until new techniques allow researchers to interpret their meaning. Elsewhere, across Europe, on alpine slopes there must be hidden caves that tell of the first stirrings of artistic impression, of extinctions and doom-filled winters.[6] "It's important to keep the collections and keep going back to them," says Joordens. "I'm sure that almost all museum collections are underexplored."

The Dubois holdings occupy an entire floor of the Naturalis Biodiversity Center and include about forty thousand specimens— cardboard boxes on shelves stretching from floor to ceiling. Within the collection the shell seems to be unique. Joordens has checked and checked again all the other shells in the collection, hoping to find more intentional markings. There are none. "It's so fragile," she says. "The shells have been slightly eroded, so the outer layer where the marked surface was has been taken off by the weathering, by being buried in the soil for such a long time." If there were marks on other shells, they're gone now.

"That we find these faint traces on this one shell—it's just a pure stroke of luck that they were carved deep enough to last."

Epilogue

In the summer of 2011, Howard Falcon-Lang, a paleontologist at Royal Holloway, University of London, was in the collections of the British Geological Survey. The low-ceilinged storage room is divided by rows of metal shelves, and on every shelf are stacks of wooden boxes with simple metal latches. Together they contain more than three million specimens collected over centuries. Falcon-Lang was walking through the collection when he noticed an incongruous-looking old wooden cabinet.

"I just pulled open the door and looked in one of the drawers," he told an interviewer in 2012.[1] Falcon-Lang reached inside. "Almost the first slide I pulled out was this one," he said, holding a microscope slide up so the light shines through it. In the middle of the slide was a darkened rectangle a little larger than a postage stamp. Etched by hand into the glass at one end: C. Darwin Esq. It was one of Charles Darwin's slides. "I could hardly believe it," he said. "My heart was pounding all around my body."

He had stumbled onto more than three hundred slides. No one had looked at them for over 160 years. Many of the slides had belonged to Joseph Hooker, a prominent botanist and a friend of Darwin's, who had worked at the British Geological Survey briefly in 1846. But a few had been Darwin's own—thin sections of fossilized wood he'd collected on Chiloé Island, off the Chilean coast. Aboard the *Beagle*, Darwin had visited the island for a few miserable rain-filled days in 1834. During his short time at the Survey, Hooker began assembling and cataloging the collection. Then in November 1847 he left on a botanical expedition to the Himalayas and never completed the task.

The slides provide a direct link to Darwin. It took a moment of serendipity to reveal them. If an untied shoelace or a buzzing cell phone had distracted Falcon-Lang he might never have found them. But he did. People make discoveries like this all the time: the frog in the jar since 1930; the tapir Roosevelt helped to collect in 1914; the unnoticed nematode worm. They have spent decades, even centuries, in obscurity. Obscurity: from the Latin *obscurus*, meaning "hidden, concealed, unknown."

This is how it happens. In the United States alone, there are an estimated one billion natural history specimens stored in biorepositories. Across the world that number probably exceeds three billion. Most of the estimated eighteen hundred natural history collections in the United States are not very big—maybe a few thousand bony fish, or a single cabinet of longicorn beetles, or a small private collection of Honduran butterflies. Often they're not housed at enormous, well-curated establishments like the Smithsonian Institution or the California Academy of Sciences or the Harvard University Museum of Comparative Zoology.

The work continues: *Capelatus prykei*, a beautiful shiny black diving beetle from the Western Cape of South Africa, collected in 1954 and named in 2015;[2] a butterfly from 1904;[3] a minuscule century-old weevil; a half-forgotten box in Paris, filled with Afro-tropical wasps collected in the 1930s and never named;[4] unidentified deep-sea bamboo worms collected from the South China Sea in 1959;[5] a new species of whirligig beetle, named from eleven specimens collected in 1974.[6] In 2016 Scripps Institution of Oceanography marine biologist Philip Hastings described *Gobiesox lanceolatus*. The canyon clingfish. The holotype — and still the only known specimen — was collected in 1965 by scientists aboard the *Soucoupe*, a cramped submersible invented by Jacques Cousteau for exploring the deep-sea environment. Usually found in shallow intertidal waters, clingfish have broad heads, stubby, tapered bodies, and an abrupt tail. This one-like a little dun-colored comma-was brought to the surface attached to a rock from the bottom of the Los Frailes canyon in Baja California, a deeper habitat than any other member of its genus. They all waited to become known. In 2009 American Museum of Natural History lepidopterist Jim Miller described *Pseudoricia flavizoma*, a strikingly pretty moth. It has powdery drab olive-gray forewings, but its hind wings are the color of a ripe lemon. It was collected in Costa Rica in March 1933. Miller found two specimens among the unsorted moths in a box at the Natural History Museum in London. Kristofer Helgen says he's aware of at least fifty unknown species of mammals waiting in museum collections. In January 2017, Helgen was part of the team that named the Skywalker hoolock gibbon (*Hoolock tianxing*), found in the tropical forests of Myanmar and southwestern China. The holotype was collected in 1917 by Roy Chapman Andrews and Yvette Borup Andrews during the American Museum of Natural History Asiatic Zoological Expedition.

Once again it raises the central question: Does it matter?

It matters.

A small nameless Costa Rican moth with luminous yellow hind wings matters. So do unknown deep-sea bamboo worms, and lungless salamanders, and Chinese gibbons that have become endangered before they were even described. Each newly named species is another glimmer in the darkness. In 1758, when Linnaeus published the tenth — and definitive — edition of *Systema Naturae*, he named seven species of bats. There are now more than twelve hundred recognized bat species. And they all tell something profound and important about the ecosystems they inhabit and how those systems are changing over time.

We humans are hard-wired to categorize the world around us — it's innate. We classify. The Stone Age artists who daubed the walls at Lascaux with running deer and bison thirty thousand years ago were early taxonomists. So were the hardworking *Homo erectus* that gathered shellfish from the water in Java 500,000 years ago, carving their zigzag patterns across the wet shells. They would have understood the subtle taxonomic differences between the eleven species of freshwater mollusks that Eugene Dubois excavated from the dig site half a million years later.

Scientists are able to describe species with long shelf lives only because the specimens have been carefully maintained in collections. But collections everywhere are threatened — by neglect, by lack of funding, by dwindling staff, by wars, and by the notion that a high-resolution image of an object is a substitute for the object itself. An infestation of dermestid beetles can devastate a collection, eating holes through irreplaceable material until it is destroyed. In the tropics the humidity invades everything, degrading specimens until they rot. In March 2011 a magnitude 9 earthquake badly damaged numerous museum collections along the east coast of Japan. In the weeks afterward, curators managed an army of volunteers who cleaned mud by hand from thousands of insect specimens, their pins bent by the force of the quake.

A 2014 paper written by thirty Italian researchers claimed the country's natural history collections had been neglected to the point of total collapse.[7] In Venezuela, the Museo Estación Biológica de Rancho Grande collection, in the city of El Limón, houses 127,000 specimens, including fifty invaluable holotypes. Owing to current political and economic upheavals, the eleven-person team that once maintained the collection has been reduced to two overworked staff members.[8] In April 2016 a fire destroyed the Museum of Natural History in New Delhi, India. Overnight, the flames burned through the six-story building, which housed India's first and only natural history collection.[9]

"The loss cannot be counted in rupees," Prakash Javadekar, India's minister of Environment, Forests, and Climate Change, told the *Times* of India. The Indian collection will be rebuilt. Curators and collection managers and field biologists will oversee the process. Even if it takes decades, they will slowly build something meaningful from the ashes.

Taxonomy is a lonely pursuit made lonelier still because fewer people are doing it now. Increasingly since 2003, DNA bar coding has begun to replace traditional taxonomy. This has ignited an unresolved debate between classical taxonomists and molecular biologists. A rapid and accurate technique, bar coding relies on small differences in the DNA sequences of different organisms to tell species apart. Undoubtedly it's a powerful tool. In August 2015 the International Barcode of Life project announced that it had met its principal goal: to bar code five million specimens representing 500,000 different species. Eventually a bar code will exist for every known species. Then, researchers will be able to isolate DNA from a specimen and use the bar code database to identify it instantly. If biodiversity is thought of as a complex symphony—with the species as the notes, the genera as the bars, and so on—DNA bar coding has the potential to identify every single note in the sheet music. But suddenly the interrelations between the notes will vanish: the pitch, the climbing scales, the tempo, the meter. The notes will become untethered from those that precede and follow them, the slowly building movements dismantled. By itself a bar code is as meaningless as a single musical note in isolation: nothing can be inferred from it. A qualified technician can generate thousands of validated bar codes in a year and never develop an informed, deeply intellectual appreciation of the evolutionary processes behind them. It takes taxonomists to do that. In doing so, they attempt to make sense of the natural world and its evolution across millions of years. From that process we begin to understand our world.

Today, in the windowless warmth of a natural history collection somewhere—in Chicago or Berlin or in Logan, Utah—a taxonomist is carefully taking a specimen-filled drawer from a shelf. She places it gently on a countertop in the monastic quiet. For the past several years the taxonomist has immersed herself in the almost imperceptible differences that define the species within a particular genus. She has studied every bud and twig on a half-forgotten branch of the Tree of Life. Insulated from the distant crowds in the exhibition halls, she trains a light on the contents of the drawer. As she inspects the old handwritten labels she begins to see patterns emerging. And half-hidden among the century-old specimens she sees something completely new.

It's something that doesn't even have a name—at least not yet.

ILLUSTRATION CAPTIONS AND CREDITS

Figure 1. A 1951 olinguito skull at the Field Museum of Natural History in Chicago. Photo: Christopher Kemp.

Figure 2. Three *Thomasomys ucucha* specimens collected in 1903, now at the American Museum of Natural History in New York City. Photo: Angelo Soto-Centeno.

Figure 3. A tapir skull—possibly a novel species—collected during Theodore Roosevelt's 1914 Brazilian expedition. Photo: Angelo Soto-Centeno.

Figure 4. The holotype for *Pithecia capillamentosa*, later renamed *P. chrysocephala*, at the Zoologische Staatssammlung in Munich, Germany, collected by Johann Baptist Spix in French Guiana in 1817. Photo: Rudolf Gerer.

Figure 5. A *Microperoryctes aplini* specimen, collected by Ernst Mayr in New Guinea in 1928. Photo: HwaJa Goetz/Museum für Naturkunde.

Figure 6. Wallace's pike cichlid, collected in April 1924 and now at the Swedish Museum of Natural History in Stockholm. Photo: Sven Kullander and Andrea Hennyey.

Figure 7. A 1919 beachcast specimen of the ruby seadragon in the Western Australian Museum fish collection. Photo: Nikolai Tatarnic.

Figure 8. A 1938 paratype of *Thorius pulmonaris* at the Louis Agassiz Museum of Comparative Zoology at Harvard University. Photo: Joseph Martinez.

Figure 9. David Blackburn conducting fieldwork in 2006 on Mount Mbam, Cameroon. Photo: Katherine S. Blackburn.

Figure 10. The holotype of *Cyrtodactylus celatus*, collected in 1924, now at the Darwin Centre at the Natural History Museum in London. Photo: Mark O'Shea.

Figure 11. *Cynipoidea* specimens that Matthew Buffington borrowed from Logan, Utah. Photo: Matthew Buffington.

Figure 12. William Clarke-Macintyre in 1939, in Ecuador. Photo: C. H. Kennedy Odonata archives from the Insect Division, Museum of Zoology, University of Michigan.

Figure 13. The holotype of *Darwinilus sedarisi*. Photo: Mark Smith, Macroscopic Solutions, from sample provided by Max Barclay, Natural History Museum, London.

Figure 14. Archived dragonfly specimens at the Royal Museum for Central Africa in Tervuren, Belgium. Photo: K. D. Dijkstra.

Figure 15. The front half of the 1907 holotype of *Rhipidocyrtus muiri* at the National Museum of Natural History in Washington, DC. Photo: Karolyn Darrow.

Figure 16. The only known specimen of *Pseudictator kingsleyae*, collected in 1896. Photo: Pierre Juhel.

Figure 17. A *Gauromydas* specimen from the 1930s archived at the National Museum of Natural History in Washington, DC. Photo: John Deamond.

Figure 18. Chris Hamilton working on *Aphonopelma* specimens at the American Museum of Natural History in New York. Photo: Amy Skibiel.

Figure 19. Eric Hoberg conducting fieldwork on Buldir Eccentric in the Aleutian Islands in 1975. Photo: courtesy of Eric Hoberg.

Figure 20. Part of the vast mollusk collection at the Natural History Museum in London. Photo: Jon Ablett.

Figure 21. High-magnification images of *Exosphaeroma paydenae*, collected in 1873 by William Healey Dall. Photo: Adam R. Wall et al., *Zookeys* 504 (2015): 11–58.

Figure 22. A 1906 specimen of *Paralomis makarovi*. Photo: John Deamond.

Figure 23. Thomas Couvreur conducting fieldwork in a Raphia forest in Cameroon in 2016. Photo: Foulques Couvreur.

Figure 24. Elmer Riggs and Harold W. Menke in the Paleontology Laboratory of the Field Columbian Museum in Chicago, 1899. Photo CSGE03248 courtesy of Photo Archives, Field Museum of Natural History.

Figure 25. A 500,000-year-old shell with intentional markings at Leiden University. Photo: Wim Lustenhouwer.

NOTES

INTRODUCTION

1. Stylianos Chatzimanolis, "Darwin's Legacy to Rove Beetles (Coleoptera, Staphylinidae): A New Genus and a New Species, Including Materials Collected on the *Beagle*'s Voyage," *ZooKeys* 379 (February 12, 2014): 29–41.

2. Kate L. Sanders et al., "*Aipysurus mosaicus*, a New Species of Egg-Eating Sea Snake (Elapidae: Hydrophiinae), with a Redescription of *Aipysurus eydouxii* (Gray, 1849)," *Zootaxa* 3431 (2012): 1–18.

3. Daniel P. Bebber et al., "Herbaria Are a Major Source of Species Discovery," *Proceedings of the National Academy of Sciences USA* 107, no. 51 (2010): 22169–71.

4. Zoë A. Goodwin et al., "Widespread Mistaken Identity in Tropical Plant Collections," *Current Biology* 25, no. 22 (2015): R1066–67.

5. Christian Rabeling et al., "*Lenomyrmex hoelldobleri*: A New Ant Species Discovered in the Stomach of the Dendrobatid Poison Frog, *Oophaga sylvatica* (Funkhouser)," *ZooKeys* 618 (September 2016): 79–95.

6. Sara Lentati, "The Man Who Keeps Finding New Species of Shark," BBC World Service, June 24, 2015.

7. Benoît Fontaine et al., "Twenty-one Years of Shelf Life between Discovery and Description of New Species," *Current Biology* 22 (2012): R943–44.

8. Christopher Kemp, "Museums: The Endangered Dead," *Nature* 518 (February 19, 2015): 292–94.

9. Michael Roston, "A Guide to Digitized Natural History Collections, *New York Times*, October 19, 2015.

10. Jurriaan M. De Vos et al., "Estimating the Normal Background Rate of Species Extinction," *Conservation Biology* 29, no. 2 (2015): 452–62.

CHAPTER ONE

1. Kristofer M. Helgen et al., "Taxonomic Revision of the Olingos (*Bassaricyon*), with Description of a New Species, the Olinguito," *ZooKeys* 324 (August 15, 2013): 1–83.

2. Julian Fennessy et al., "Multi-locus Analyses Reveal Four Giraffe Species Instead of One," *Current Biology* 26, no. 18 (2016): 2543–49.

3. Joseph Stromberg, "For the First Time in 35 Years, a New Carnivorous Mammal Species Is Discovered in the Americas." Smithsonian.com, August 14, 2014.

CHAPTER TWO

1. Robert S. Voss, "A New Species of *Thomasomys* (Rodentia: Muridae) from Eastern Ecuador, with Remarks on Mammalian Diversity and Biogeography in the Cordillera Oriental," *American Museum Novitates*, no. 3421, 2003.

2. Samuel N. Rhoads, "Letter from Hacienda Gorzon, South Foot of Mt. Pinchacha, Six Miles from Quito, Fifth Mo. 16th," *Friend: A Religious and Literary Journal* 84–85 (1911): 30.

CHAPTER THREE

1. Mario A. Cozzuol et al., "A New Species of Tapir from the Amazon," *Journal of Mammalogy* 94, no. 6 (2013): 1331–45.

2. Theodore Roosevelt, *Through the Brazilian Wilderness* (New York: Charles Scribner's Sons, 1914).

3. Joel A. Allen, "Mammals Collected on the Roosevelt Brazilian Expedition, with Field Notes by Leo E. Miller," *Bulletin of the American Museum of Natural History* 35 (1914): 559–610.

4. Izeni P. Farias et al., "The Cytochrome b Gene as a Phylogenetic Marker: The Limits of Resolution for Analyzing Relationships among Cichlid Fishes." *Journal of Molecular Evolution* 53 (2001): 89–103.

5. Robert S. Voss et al., "Extraordinary Claims Require Extraordinary Evidence: A Comment on Cozzuol et al. (2013)," *Journal of Mammalogy* 95, no. 4 (2014): 893–98.

6. Kristofer M. Helgen et al., "Pacific Flying Foxes (Mammalia: Chiroptera): Two New Species of *Pteropus* from Samoa, Probably Extinct," *American Museum Novitates*, no. 3646, 2009.

7. Mario A. Cozzuol et al., "How Much Evidence Is Enough Evidence for a New Species?" *Journal of Mammalogy* 95, no. 4 (2014): 899–905.

8. Ronald H. Pine et al., "A New Species of the Didelphid Marsupial Genus *Monodelphis* from Eastern Bolivia," *American Museum Novitates*, no. 3740, 2013.

9. Trevor Jones et al., "The Highland Mangabey *Lophocebus kipunji*: A New Species of African Monkey," *Science* 308 (May 2005): 1161–64.

10. Stephen Marshall et al., "New Species without Dead Bodies: A Case for Photo-Based Descriptions, Illustrated by a Striking New Species of *Marleyimyia* Hesse (Diptera, Bombyliidae) from South Africa," *ZooKeys* 525 (October 5, 2015): 117–27.

11. Evan Ratliff, "Why Does This Prominent Amazon Researcher Face 14 Years in Prison for Biopiracy?" *Wired*, May 19, 2008.

12. International Commission on Zoological Nomenclature, "Case 3650: *Tapirus pygmaeus* Van Roosmalen and Van Hooft in Van Roosmalen, 2013 (Mammalia, Perissodactyla, Tapiridae): Proposed Confirmation of Availability of the Specific Name and of the Book in Which This Nominal Species Was Proposed" (2014).

CHAPTER FOUR

1. Laura K. Marsh, "A Taxonomic Revision of the Saki Monkeys, *Pithecia* Desmarest, 1804," *Neotropical Primates* 21, no. 1 (2014): 1–165.

2. Georges L. L. Buffon, *Natural History, General and Particular* (London: Strahan and Cadell, 1785).

3. Alfred Russel Wallace, "On the Monkeys of the Amazon," *Proceedings of the Zoological Society of London*, 1854.

4. Philip Hershkovitz, "The Taxonomy of South American Sakis, Genus *Pithecia* (Cebidae, Platyrrhini): A Preliminary Report and Critical Review with the Description of a New Species and a New Subspecies," American *Journal of Primatology* 12 (1987): 387–468.

CHAPTER FIVE

1. Christian Thompson, Final Frontier: Newly Discovered Species of New Guinea (1998–2008), World Wildlife Fund Report, 2011.

2. "Lost World of Fanged Frogs and Giant Rats Discovered in Papua New Guinea," *Guardian*, September 6, 2009.

3. Alfred Russel Wallace, *My Life: A Record of Events and Opinions* (London: Chapman and Hall, 1905).

4. Ernst Mayr, "My Dutch New Guinea Expedition, 1928," *Novitates Zoologicae: A Journal of Zoology* 36, no. 489 (1930–31): 1–10.

5. M. A. Salmon et al., *The Aurelian Legacy: British Butterflies and Their Collectors* (Berkeley: University of California Press, 2000).

6. L. M. D'Albertis, "Journey to the Arfak Mountains, New Guinea," *Melbourne Review*, no. 3, July 1876.

7. George Ockenden Obituary, *Novitates Zoologicae: A Journal of Zoology* 14, no. 1 (1908): 341–42.

8. Antwerp E. Pratt, Two Years among New Guinea Cannibals: A Natural-

ist's Sojourn among the Aborigines of Unexplored New Guinea (Philadelphia: J. B. Lippincott, 1906).

9. Interview with Ernst Mayr, "Web of Stories: From Dutch New Guinea to the Mandated Territory of New Guinea," 1997.

10. Kristofer M. Helgen et al., "A New Species of Bandicoot, *Microperoryctes aplini*, from Western New Guinea," *Journal of Zoology* 264, no. 2 (2004): 117–24.

11. Michael Westerman et al., "Phylogenetic Relationships of Living and Recently Extinct Bandicoots Based on Nuclear and Mitochondrial DNA Sequences," *Molecular Phylogenetics and Evolution* 62 (2012): 97–108.

CHAPTER SIX

1. Alfred Russel Wallace, *My Life: A Record of Events and Opinion* (London: Chapman and Hall, 1905).

2. Sven O. Kullander et al., "Wallace's Pike Cichlid Gets a Name after 160 Years: A New Species of Cichlid Fish (Teleostei: Cichlidae) from the Upper Rio Negro in Brazil," *Copeia* 103, no. 3 (2015): 512–19.

CHAPTER SEVEN

1. "Rare Ruby Seadragon Washes Up on WA Cliffs," *Australian Geographic*, March 5, 2015.

2. Josefin Stiller, Nerida G. Wilson, and Greg W. Rouse, "A Spectacular New Species of Seadragon (Syngnathidae)," *Royal Society Open Science* 2 (2015): 140458.

3. Julian Edmund Tenison-Woods, *Fish and Fisheries of New South Wales* (Sydney: T. Richards, 1882).

CHAPTER EIGHT

1. James Hanken et al., "Biology of the Tiny Animals: Systematics of the Minute Salamanders (*Thorius*: Plethodontidae) from Veracruz and Puebla, Mexico, with Descriptions of Five New Species," *Copeia* 1998 (2): 312–45.

2. Hans Gadow, Through Southern Mexico, Being an Account of the Travels of a Naturalist (London: Witherby, 1908).

3. Sean M. Rovito et al., "Adaptive Radiation in Miniature: The Minute Salamanders of the Mexican Highlands (Amphibia: Plethodontidae: *Thorius*)," *Biological Journal of the Linnean Society* 109 (2013): 622–43.

4. Gabriela Parra-Olea et al., "Biology of Tiny Animals: Three New Species of Minute Salamanders (Plethodontidae: *Thorius*) from Oaxaca, Mexico," *PeerJ* 4 (2016):e2694, doi: 10.7717/peerj.2694.

CHAPTER NINE

1. David C. Blackburn, "New Species of *Arthroleptis* (Anura: Arthroleptidae) from Ngozi Crater in the Poroto Mountains of Southwestern Tanzania," *Journal of Herpetology* 46, no. 1 (2012): 129–35.

2. Mark O'Shea et al., "A New Species of New Guinea Worm-Eating Snake, Genus *Toxicocalamus* (Serpentes: Elapidae), from the Star Mountains of Western Province, Papua New Guinea, with a Revised Dichotomous Key to the Genus," *Bulletin of the Museum of Comparative Zoology* 161, no. 6 (2015): 241–64.

3. J. A. Feinberg et al., "Cryptic Diversity in Metropolis: Confirmation of a New Leopard Frog Species (Anura: Ranidae) from New York City and Surrounding Atlantic Coast Regions," *PLoS One* 9, no. 10 (2014): e108213, doi: 10.1371/journal.pone.0108213.

4. Arthur Loveridge, "Reports on the Scientific Results of an Expedition to the Southwestern Highlands of Tanganyika Territory. VII. Herpetology," *Bulletin of the Museum of Comparative Zoology at Harvard University* 74, no. 7 (1933): 195–416.

5. Joseph R. Mendelson, "Shifted Baselines, Forensic Taxonomy, and Rabbs' Fringe-Limbed Treefrog: The Changing Role of Biologists in an Era of Amphibian Declines and Extinctions, *Herpetological Review* 42, no. 1 (2011): 21–25.

6. John Upton, "Despite Deadly Fungus, Frog Imports Continue," *New York Times*, April 7, 2012.

CHAPTER TEN

1. Malcolm A. Smith, "Contributions to the Herpetology of the Indo-Australian Region," *Proceedings of the Zoological Society* of London, 1927, part 1.

2. "An Eminent Company of Amateur Naturalists," *Life*, September 10, 1956.

3. Malcolm A. Smith, "Large Banded Krait," *Journal of the Natural History Society of Siam* 1, no. 1 (1914): 5–18.

4. Malcolm A. Smith, "The Snakes of Bangkok," *Journal of the Natural History Society of Siam* 1, no. 2 (1914): 94–104.

5. Malcolm A. Smith Obituary, *Nature*, September 13, 1958.

6. Hinrich Kaiser et al., "The Herpetofauna of Timor: A First Report," *ZooKeys* 109 (June 20, 2011): 19–86.

7. Andrew Kathriner et al., "Hiding in Plain Sight: A New Species of Bent-Toed Gecko (Squamata: Gekkonidae: *Cyrtodactylus*) from West Timor, Collected by Malcolm Smith in 1924," *Zootaxa* 3900, no. 4 (2014): 555–68.

8. Malcolm A. Smith, "A Bangkok Python," *Natural History Bulletin of the Siam Society* 11, no. 1 (1937): 61–62.

CHAPTER ELEVEN

1. Simon Van Noort, Matthew L. Buffington, and Mattias Forshage, "Afrotropical Cynipoidea (Hymenoptera)," *ZooKeys* 493 (April 1, 2015): 1–176.

2. R. Brookes, The Natural History of Insects, with Their Properties and Uses in Medicine, vol. 14 (London: Newberry, 1763).

3. Katherine N. Schick and Donald L. Dahlsten, "Gallmaking and Insects," in *Encyclopedia of Insects*, ed. V. H. Resh and R. Cardé (New York: Academic Press, 2003).

CHAPTER TWELVE

1. Peter Vršanský et al., "Light-Mimicking Cockroaches Indicate Tertiary Origin of Recent Terrestrial Luminescence," *Naturwissenschaften* (2012), doi: 10.1007/s00114–012–0956–7.

2. Oliver Zompro et al., "*Lucihormetica* n. gen. n. sp., the First Record of Luminescence in an Orthopteroid Insect (Dictyoptera: Blaberidae: Blaberinae: Brachycolini), *Amazoniana* 15 (1999): 211–19.

3. Clark-Macintyre correspondence, C. H. Kennedy Odonata archives from the Insect Division, Museum of Zoology, University of Michigan.

4. Charles P. Alexander, "Notes on the Tipulidae of Ecuador," *Revista Ecuatoriana de Entomologia y Parasitologica* 1, no. 1 (1953): 66.

5. Clarence H. Kennedy, "Odonata Archives from the Insect Division, Museum of Zoology, University of Michigan (1936–1950)."

6. BBC.com, "Tungurahua Volcano Prompts Evacuation in Ecuador," December 4, 2010.

CHAPTER THIRTEEN

1. Charles Darwin, *The Voyage of the "Beagle"* (London: Henry Colburn, 1839).

2. Charles Darwin, The Autobiography of Charles Darwin, 1809–1882, with the Original Omissions Restored, Edited and with Appendix and Notes by His Grand-daughter Nora Barlow (London: Collins, 1958).

3. K. G. V. Smith, ed., "Darwin's Insects: Charles Darwin's Entomological Notes," *Bulletin of the British Museum of Natural History (Historical Series)* 14, no. 1 (1987): 1–143.

4. Stylianos Chatzimanolis, "Darwin's Legacy to Rove Beetles (Coleoptera, Staphylinidae): A New Genus and a New Species, Including Materials Collected on the *Beagle*'s Voyage," *ZooKeys* 379 (February 12, 2014): 29–41.

CHAPTER FOURTEEN

1. Adam Hochschild, King Leopold's Ghost: A Story of Greed, Terror, and Heroism in Colonial Africa (Boston: Houghton Mifflin, 1998).

2. Klaas-Douwe B. Dijkstra, Jens Kipping, and Nicolas Mézière, "Sixty New Dragonfly and Damselfly Species from Africa (Odonata)," *Odontologica* 44, no. 4 (2015): 447–678.

3. J. W. H. Trueman et al., "Odonata: Dragonflies and damselflies," Tree of Life Web Project, 2009. http://tolweb.org/Odonata.

CHAPTER FIFTEEN

1. Frederick Muir, "Muir's Report of Travels in Borneo in Search of Cane Borer Parasites," *Hawaiian Planters' Monthly*, February 1908.

2. Zachary H. Falin and Michael S. Engel, "Serendipity at the Smithsonian: The 107-Year Journey of *Rhipidocyrtus muiri* Falin and Engel, New Genus and

Species (Ripidiinae, Ripidiini), from Jungle Beats to Valid Taxon," *ZooKeys* 424 (June 8, 2014): 101–16.

3. C. V. Riley, "A Sandwich Islands Sugar-Cane Borer," *Insect Life* 1, no. 6 (1888): 185–89.

4. Leland O. Howard, "On the Hawaiian Work in Introducing Beneficial Insects," *Hawaiian Planters' Record*, June/July, 1916.

5. Paul DeBach and David Rosen, *Biological Control by Natural Enemies*, 2nd ed. (London: Cambridge University Press, 1991).

CHAPTER SIXTEEN

1. Mary Kingsley, *Travels in West Africa* (New York: Macmillan, 1897).

2. "Dr. Livingstone's I Presume? Natural History Museum Finds Explorer's African Insect Collection," *Independent*, September 19, 2014.

3. Katherine Frank, *A Voyage Out: The Life of Mary Kingsley* (Boston: Houghton Mifflin, 1986).

CHAPTER SEVENTEEN

1. Bruna Klassa et al., "The Man Who Loved Flies: A Biographical Profile of Nelson Papavero," *Zootaxa* 3793, no. 2 (2014): 201–21.

2. Joseph R. Coelho et al., "The Effect of Hind-Tibial Spurs on Digging Rate in Female Eastern Cicada Killers," *Ecological Entomology* 33, no. 3 (2008): 403–7.

3. Julia Calhau, C. J. E. Lamas, and S. S. Nihei, "Review of the *Gauromydas* Giant Flies (Insecta, Diptera, Mydidae), with Descriptions of Two New Species from Central and South America," *Zootaxa* 4048, no. 3 (2015): 392–411.

4. Justin O. Schmidt, "Venom and the Good Life in Tarantula Hawks (Hymenoptera: Pompilidae): How to Eat, Not Be Eaten, and Live Long," *Journal of the Kansas Entomological Society* 77, no. 4 (2004): 402–13; Justin O. Schmidt, "Hymenoptera Venoms: Striving toward the Ultimate Defense against Vertebrates," in *Insect Defenses: Adaptive Mechanisms and Strategies of Prey and Predators*, ed. D. L. Evans and J. O. Schmidt, 387–419 (Albany: State University of New York Press, 1990).

CHAPTER EIGHTEEN

1. Chris A. Hamilton et al., "Taxonomic Revision of the Tarantula Genus *Aphonopelma* Pocock, 1901 (Araneae, Mygalomorphae, Theraphosidae) within the United States," *ZooKeys* 560 (February 4, 2016): 1–340.

2. W. J. Gertsch, *American Spiders* (New York: Van Nostrand, 1949).

3. "Fourteen New Tarantula Species Found in United States," *Newsweek*, February 4, 2016.

4. Andrew Smith, Tarantula Spiders: Tarantulas of the USA and Mexico (London: Fitzgerald, 1994).

CHAPTER NINETEEN

1. Marie-Claude Durette-Desset, Kurt E. Galbreath, and Eric P. Hoberg, "Discovery of New *Ohbayashinema* spp. (Nematoda: Heligmosomoidea) in *Ochotona princeps* and *Ochotona cansus* (Lagomorpha: Ochotonidae) from Western North America and Central Asia, with Considerations of Historical Biogeography," *Journal of Parasitology* 96, no. 3 (2010): 569–79.

2. Nathan Augustus Cobb, *Nematodes and Their Relationships*, Yearbook of the Department of Agriculture (Washington, DC: US Department of Agriculture, 1914).

CHAPTER TWENTY

1. Barna Páll-Gergely et al., "Revision of the Genus *Pseudopomatias* and Its Relatives (Gastropoda: Cyclophoroidea: Pupinidae)," *Zootaxa* 3937, no. 1 (2015): 1–49.

2. Maurice Isserman and Stewart Weaver, Fallen Giants: A History of Himalayan Mountaineering from the Age of Empire to the Age of Extremes (New Haven, CT: Yale University Press, 2010).

3. Fred Naggs, "William Benson and the Early Study of Land Snails in British India and Ceylon," *Archives of Natural History* 24, no. 1 (1997): 37–88.

CHAPTER TWENTY-ONE

1. Douglas Main, "New Aquatic 'Roly-Poly' Found in Los Angeles," *Newsweek*, May 18, 2015.

2. Adam R. Wall et al., "Status of *Exosphaeroma amplicauda* (Stimpson, 1857), *E. aphrodita* (Boone, 1923) and Description of Three New Species (Crustacea, Isopoda, Sphaeromatidae) from the North-Eastern Pacific," *ZooKeys* 504 (May 18, 2015): 11–58.

3. "Thirty Never-Before-Seen Species of Flies Discovered in Los Angeles," *Los Angeles Times*, March 27, 2015.

4. E. A. Hartop et al., Opportunity in Our Ignorance: Urban Biodiversity Study Reveals 30 New Species and One New Nearctic Record for Megaselia (Diptera: Phoridae) in Los Angeles (California, USA)," *Zootaxa* 3941, no. 4 (2015): 451–84.

5. Alfred G. Mayer, *Biographical Memoir of William Stimpson*, Biographical Memoirs, vol. 3 (Washington, DC: National Academy of Sciences of the United States of America, 1918).

6. William Healy Dall, Field Notes, Smithsonian Transcription Project, 1873.

CHAPTER TWENTY-TWO

1. Arthur Hobart Clark, Dredging and Hydrographic Records of the U.S. Fisheries Steamer "Albatross" for 1906 (Washington, DC: United States Bureau of Fisheries, 1907).

2. Sally Hall and Sven Thatje, "Four New Species of the Family Lithodidae

(Decapoda: Anomura) from the Collections of the National Museum of Natural History, Smithsonian Institution," *Zootaxa* 2302 (2009): 31–47.

3. Mark Jennings, "The U.S. Fish Commission Steamer *Albatross*: A History," *Marine Fisheries Review* 61, no. 4 (1999): i–vii.

4. Austin Hobart Clark, "The Birds Collected and Observed during the Cruise of the United States Fisheries Steamer *Albatross* in the North Pacific Ocean, and in the Bering, Okhotsk, Japan and Eastern Seas, from April to December, 1906," *Proceedings of the United States National Museum* 38, no. 1727 (1906): 1–74.

5. J. Richard Dunn, "Charles Henry Gilbert (1859–1928), Naturalist-in-Charge: The 1906 North Pacific Expedition of the Steamer *Albatross*," *Marine Fisheries Review* 58, nos. 1–2 (1996): 17–28.

6. Quentin D. Wheeler, "Taxonomic Triage and the Poverty of Phylogeny," *Philosophical Transactions of the Royal Society of London B, Biological Sciences* 359, no. 1444 (2004): 571–83.

CHAPTER TWENTY-THREE

1. Paul H. Hoekstra et al., "A Nonet of Novel Species of *Monanthotaxis* (Annonaceae) from around Africa," *PhytoKeys* 69 (2016): 71–103.

2. George A. McCue, "The History of the Use of the Tomato: An Annotated Bibliography," *Annals of the Missouri Botanical Gardens* 39, no. 4 (1952): 289–348.

3. Daniel P. Bebber et al., "Herbaria Are a Major Frontier for Species Discovery," *Proceedings of the National Academy of Sciences USA* 107, no. 51 (2010): 22169–71.

CHAPTER TWENTY-FOUR

1. Mark Jaffe, The Gilded Dinosaur: The Fossil War between E. D. Cope and O. C. Marsh and the Rise of American Science (New York: Crown, 2000).

2. Susumu Tomiya et al., "Whence the Beardogs? Reappraisal of the Middle to Late Eocene 'Miacis' from Texas, USA, and the Origin of Amphicyonidae (Mammalia, Carnivora)," *Royal Society Open Science* 3 (2016): 160518.

3. Michael Hanson and Peter J. Makovicky, "A New Specimen of *Torvosaurus tanneri* Originally Collected by Elmer Riggs," *Historical Biology: An International Journal of Paleobiology* 26, no. 6 (2014): 775–84.

4. Peter M. Galton and James A. Jensen, "A New Large Theropod Dinosaur from the Upper Jurassic of Colorado," *Brigham Young University Geology Studies* 26, no. 1 (1979): 1–12.

5. Elmer S. Riggs, *Mounted Skeleton of "Homalodotherium,"* Geological series, vol. 6, no. 17 (Chicago: Field Museum of Natural History, 1909).

6. Bruce J. Shockey et al., "New Leontiniid Notoungulata (Mammalia) from Chile and Argentina: Comparative Anatomy, Character Analysis, and Phylogenetic Hypotheses," *American Museum Novitates*, no. 3737, 2012.

CHAPTER TWENTY-FIVE

1. Thomas Sutikna et al., "Revised Stratigraphy and Chronology for *Homo floresiensis* at Liang Bua in Indonesia," *Nature* 532 (April 21, 2016): 366–69.

2. "*Homo erectus* Made World's Oldest Doodle 500,000 Years Ago," *Nature*, December 3, 2014.

3. Jose C. A. Joorden et al., "*Homo erectus* at Trinil on Java Used Shells for Tool Production and Engraving," *Nature* 518 (February 12, 2015): 228–31.

4. Peter Brown et al., "A New Small-Bodied Hominin from the Late Pleistocene of Flores, Indonesia," *Nature* 431 (October 28, 2004): 1055–61.

5. Johannes Krause et al., "The Complete Mitochondrial DNA Genome of an Unknown Hominin from Southern Siberia," *Nature* 464 (April 8, 2010): 894–97.

6. Joaquín Rodriguez-Vidal et al., "A Rock Engraving Made by Neanderthals in Gibraltar," *Proceedings of the National Academy of Sciences* 111, no. 37 (2014): 13301–6.

EPILOGUE

1. Howard Falcon-Lang, "Fossil 'Treasure Trove' Found in British Geological Survey Vaults," *Geology Today* 28, no. 1 (2012): 26–30.

2. David T. Bilton et al., "*Capelatus prykei* gen. et sp. n. (Coleoptera: Dytiscidae: Copelatinae)—a Phylogenetically Isolated Diving Beetle from the Western Cape of South Africa," *Systematic Entomology* 40 (2015): 520–31.

3. Blanca Huertas, "A New Species of Satyrinae Butterfly from Peru (Nymphalidae: Satyrini: Euptychiina)," *Zootaxa* 2802 (March 2002): 63–68.

4. Simon Van Noort et al., "Revision of the Afrotropical Mayrellinae (Cynipoidea, Liopteridae), with the First Record of Paramblynotus from Madagascar," *Journal of Hymenoptera Research* 31 (2013): 1–64.

5. Y. Wang, et al., "A New Maldane Species and a New Maldaninae Genus and Species (Maldanidae, Annelida) from Coastal Waters of China," *ZooKeys* 603 (July 6, 2016): 1–16.

6. Grey T. Gustafson et al., "A North American Biodiversity Hotspot Gets Richer: A New Species of Whirligig Beetle (Coleoptera: Gyrinidae) from the Southeastern Coastal Plain of the United States," *Annals of the Entomological Society of America* 109, no. 1 (2016): 42–48.

7. Franco Andreone et al., "Italian Natural History Museums on the Verge of Collapse?" *ZooKeys* 456 (November 24, 2014): 139–46.

8. Eliécer C. Gutiérrez et al., "Venezuelan Crisis Takes Toll on Natural History Museum," *Herpetological Review* 47, no. 4 (2016): 710–11.

9. "India Fire Destroys Natural History Museum in Delhi," BBC.com, April 26, 2016.

INDEX

Ablett, Jon, 170–71, 175, 224
Ablett's land snail, 169
Academy of Natural Sciences of
    Drexel University, 27
Acaronia vultuosa, 55
Acre State, Brazil, 37
Adelaide Environment Institute, 3
Africa, 53, 78, 120–21, 124, 137, 196–
    97, 211, 213–14
African dragonflies, 120, 122, 175
African lake cichlid species, 53
African longhorn beetles, 135, 137
Afrotropical wasps, 218
Afzelius, Adam, 198
Aipysurus mosaicus, xv
Alaska, 165–66, 181, 185, 188
Alberta, Canada, 202
Aleutian island chain, 165, 181–82, 185,
    189, 224
Allen, Joel Asaph, 23
Allosaurus, 203
Alto Rio Juruá, Brazil, 37

Amazon, 14, 19–23, 26, 28–29, 31, 33–
    37, 42, 51–52, 54–55, 57, 105,
    158–29
Amazonas State, Brazil, 21, 23, 28
Ambon, Indonesia, 131
Amchitka, Alaska, 182
American Museum of Natural His-
    tory, xvii, xxi, 6, 13, 15–17, 21–
    23, 32, 36, 44, 48, 152–53, 203,
    207, 219, 223–24
American Spiders, Gertsch, 153
amphibians, 5, 43, 75, 78, 86, 90, 106,
    112–31
amphicyonids, 205
Anax gladiator, 123
Andes, 6, 8, 17, 36, 52, 81, 104
Andricus foecundatrix, 98
Anggi lakes, New Guinea, 49
Angola, 79, 97, 120
Annonaceae family, 194
Anomala orientalis, 132
Antarctica, xxi, 121, 163, 165

Anthony, Harold Elmer, 16
Aphonopelma atomicum, 151–53, 157
Aphonopelma catalina, 158–31
Aphonopelma chamberlini, 154
Aphonopelma coloradanum, 154
Aphonopelma johnnycashi, 155
aphotic zone, 63, 65
Aplin, Ken, 47
Archbold field expeditions, 77
Arctic, 164–67
Arfak Mountains, New Guinea, 42–44, 47–48
Arfak Pygmy Bandicoot, 41, 46, 48–49
Argentina, 25, 100, 111, 115, 202, 206–7
Arizona, 95, 101, 154–55, 158
Armagosa Desert, Nevada, 153
Armbruster, Jonathan, 57
Arthroleptis adolfifriederici, 75–76, 79
Arthroleptis aureoli, 76
Arthroleptis kutogundua, 75–76, 78, 80–82
Arthroleptis nimbaensis, 76
Arthroleptis stenodactylus, 82
Arthroleptis troglodytes, 76
Ascaris columnaris, 162
Asia, 53–54, 87, 170, 172, 211
Asian land snails, 173
Astrolabe, French ship, 131
Atelopus genus, 81
Atlanta Zoo, 81
Atomic Tarantula Spider, 151
atomic tests, 51, 54, 151
Australia, 41–42, 45, 47, 54, 61, 64, 87, 131–32, 146, 180, 212
Australian National Fish Collection, 63
autotomy, 88

Badlands, South Dakota, 203
Bahía Blanca, 111, 113, 115

Baja California, 219
Baker, E. C. Stuart, 47
Bakewell's thread snake, 69
bald-faced saki, 37
Bali, 91
bandicoots, 41, 47–49
Bangkok, 87
Baños in Tungurahua Province, 104–7
Barber, Herbert, 132
Barclay, Max, xvii, 135–40, 224
bar coding, 24, 221
Bassaricyon genus, 4–5, 7–8
Bassaricyon neblina, 3
Bassaricyon neblina hershkovitzi, 7
Bassaricyon neblina neblina, 6
Bathynomus giganteus, 180
Bathyuroconger parvibranchialis, 189
Batrachochytrium dendrobatidis fungus, 81
bats, xvi–xvii, xxi, 3, 8, 24, 42, 46, 105, 219
Bavarian State Collection of Zoology, Munich, 56
beardogs, 205
Bee Lab, 96–97, 101, 105, 147, 179
beetles, xiv, xvii, xx, 4, 43, 95, 104, 111–15, 128–29, 131–33, 135–40, 145, 147, 218, 220
    ripidiine, 130, 132
    toxic *Pyrophorus*, 104
    wedge-shaped, 128–29, 133
    whirligig, 218
Belgian Congo, 119–20
Belgium, 119–20, 211, 224
Belo Horizonte, Brazil, 19
Benson, William Henry, 172–73
Bent-Toed Gecko, 88, 90
Bering Sea, 166, 181, 185
Berlin, Germany, xix–xx, 34, 47, 68, 81, 90, 96, 114, 221
Bernice P. Bishop Museum, Honolulu, xvii, 47

Bicyclus genus, 124

biodiversity, xiv, xvi, 8, 14, 24, 31, 72, 120, 124, 130, 133, 175, 179, 182, 190

biomimicry, 103, 105, 107

BioScan, 179

birds, 15–16, 37, 42, 45–46, 54, 104–5, 112, 138, 143, 147, 164, 166, 196, 206–7

Bird's Head Peninsula, New Guinea, 41–42

birds-of-paradise, 44–45

Bird's Tail Peninsula, Papua New Guinea, 42

Blackburn, David C., 75–82, 124, 224

Blanford, WT, 170, 172–73

Blattodea collection, 103–4

Boer War, 140

Bogor Museum in Java, 47

Boisduval, Jean Baptiste, 131

Boissonneaua jardini, 105

Bolivia, 22, 96, 202, 206–7

bones, 4–5, 71, 169, 187, 201–2, 204–7, 212, 214

Bone Wars, 202

Bonn, Germany, 186

Borneo, xxi, 54, 127–33

Boschniakia glabra, 181

botanical specimens, xvii, 69, 155, 158, 191, 195

botanists, 113, 193–94, 196–97, 217

Bowers Bank, Alaska, 185–87

Brachiosaurus, 202–3

Brachyura, 187

Brazil, 19, 21, 29, 32–35, 37, 52, 55, 143, 145–46, 148, 157–58

British Geological Survey, 217–18

British Museum, 15–16, 44–45, 54, 86, 100, 169

Brontosaurus, 203, 206

Bruce, Niel, 180

Brussels, Belgium, 96, 119–20

Bryant, Owen, 155

Buffington, Matthew L., 95–101, 128, 224

Buffon, Georges Louis Leclerc, 33

Bungarus fasciatus, 87

butterflies, 43, 47, 69, 124, 179

Bynesian decay, 173

Caenolestes fuliginosus, 16

Calhau, Julia, 143–45, 147–48

California, 68, 80, 86, 153–54, 178–79

California Academy of Sciences, xvii, 95–96, 101, 148, 155, 218

Callirhytis cornigera, 98

Callosobruchus maculatus, 146

Camarasaurus, 203

Cambridge University, 136

Cameroon, 79, 81, 120, 140, 193, 195–97, 224–25

Campbell, Patrick, 86, 88

Canada, 164–65, 202–3

Canadian Northwest Territories, 165

Cane Borer Parasites, 128, 130–32

Capelatus prykei, 218

Cape Town, South Africa, 96, 120, 140

Cash, Johnny, 155

catfish, 56–57

Central Africa, 119–20, 122, 198, 224

Central America, 81, 145, 154

Central Cordillera of New Guinea, 49

Cerambycidae subfamily of beetles, 137

Chamberlin, Ralph V., 154

Chatzimanolis, xiv, xvii, 112–15

Chicago, xvii, 3, 27, 180, 201, 208, 221, 223, 225

Chihuahuan Desert, 154

Chiloé Island, 218

China, 127, 131, 174–75, 219

Chiricahua Mountains, 154–55

Chlorocypha granata, 123

Chordata phylum, 112

Chrysis marqueti, xx
chytridiomycosis, 81
Cibinong, Indonesia, 47
cichlids, 52–53, 55–57
Clarke-Macintyre, William, 104–8
Cobb, Nathan Augustus, 164
cockroach, 103–5, 107–8, 129, 132
    lightning, 103–4, 106–8, 146
Coelophysis, 203
Coleoptera, 106, 112
Colombia, 5, 7, 25, 52, 105
Colorado, 104, 154, 158, 202–3, 206
Congo, 42, 119–24, 196
Congo Duskhawker Dragonfly, 119–20
Congo River, 121
Congrhynchus talabonoides, 189
conservationists, xiv, 27, 72, 80
Cope, Edward Drinker, 68, 202
Copenhagen, Denmark, xv, 139
Cordillera Oriental, Ecuador, 13, 17,
    105–6
Costa Rica, 145, 219
Cottesloe Beach, 59, 65
Cousteau, Jacques, 219
Couvreur, Thomas, 193–95, 197–98,
    225
Cozzuol, Mario A., 19–23, 25–26, 28–
    29
crabs, xv, 186–88, 190
Crenicichla monicae, 51–52, 55–57
Cretaceous dinosaurs, late, 203
Crustacea, 178, 182, 188
Cyclopecten davidsoni, 190
Cyclops Mountains, New Guinea, 44
cynipid wasps, 95–101, 128, 224
Cyrtodactylus celatus, 85, 88, 224
Cyrtodactylus wetariensis, 86

Dactylopsila trivergata, 46
D'Albertis, 45–46
Dall, William Healey, 181, 224
Damselfly species, 105, 120, 145

Darjeeling, 170, 172, 175
Darwin, Charles, 53–54, 111, 113–15,
    155, 217–18
Darwinilus sedarisi, xiv, 111, 115, 224
Darwin's rove beetle, 111
Davenport, Tim, 27
Death Valley, Nevada, 153
Democratic Republic of the Congo,
    119–21
Denisovans, 214
Department of Entomology, NMNH,
    100, 103–4, 108
Department of Invertebrate Zoology,
    NMNH, 186
Desmarest, 33
digitizing natural history collections,
    xix–xx, 138
Dijkstra, Klaas-Douwe, 120–25, 224
Dikmans, Gerard, 162
Dikow, Torsten, 146–48
Dimetrodon, 203
Diplodocus, 203
Diplolepis rosae, 98
Diptera, 143–45, 148–34
Diringshofen, 144
Disholcaspis plumbella, 98
Disney World, Florida, 204
Djamplong, West Timor, 88, 90
DNA analysis, 5, 9, 23–25, 27, 61–62,
    68, 157, 173–74, 196, 221
DNA bar coding, 24, 221
Doherty, William, 46
dragonflies, 105, 121–22
Dresden, Germany, 45
Drosophila busckii, 146
Dubois, Eugene, 211–12, 215, 220
duck-billed dinosaurs, 204, 207
Dumont, Jules, 131

earliest hominin engraving, 211
East Africa, 75, 80, 82
East India Company, 171–72

East Timor, 88–89
Eckhout, Albert, 158
Ecuador, xvii, 4–8, 13–16, 31, 35, 103–4, 106–8, 224
eggs, 62, 76, 97–99, 115, 128, 131
Elmerriggsia fieldia, 207
El Reventador volcano, Ecuador, 108
Ensatina genus, 80
entomology, xv, xvii, xix, 99–100, 103–4, 106–8, 112, 115, 120–21, 127–28, 130, 135, 137, 144–45, 147–48
evolution, xiv, 15, 43, 53, 63, 148, 156, 197, 221
Exosphaeroma amplicauda, 179–82
Exosphaeroma aphrodita, 181
Exosphaeroma paydenae, 177, 182, 224
Exosphaeroma pentcheffi, 178
Exosphaeroma russellhansoni, 181
extinctions, xvi, xx, 44, 81, 115, 175, 205, 207, 213, 215

Falcon-Lang, Howard, 217–18
Falin, Zachary H., 128–30, 132–33
Fernholm, Bo, 53
field biologists, xvii, 4, 13, 15, 23, 68, 78, 80, 88–89, 104, 108, 120–21, 196, 201, 205
Field Columbian Museum, 202, 225
field jackets, 202, 204–8
Field Museum of Natural History, Chicago, xvii–xix, 3–4, 7, 9, 16, 24, 27, 106, 197, 201–5, 207, 223, 225
Finnish Museum of Natural History, Helsinki, 167
fish, xv, 5, 43, 52–57, 60–61, 112, 137, 140, 164, 189, 206, 218
Flannery, Tim, 48
Florida Museum of Natural History, Gainesville, 75, 152
flowers, 47, 128, 132, 144, 193–94, 196
fly, 131–32, 138, 143–48

Fly River, Papua New Guinea, 45
fossils, 111, 180, 201–7, 211–12–36
France, 89, 193, 213
Frankfurt, Germany, 9, 90, 173, 186
Fraser, Frederic Charles, 122
Freezeout Hills, Wyoming, 205
French Guiana, 33, 223
frogs, xv–xvii, 42, 56, 69, 75–79, 81–82, 86, 90, 105, 155, 218
   squeaker, 75–76, 78, 80, 124
Fülleborn, Friedrich, 82
FUNAI, 37
Furth, David, 107–8

Gabon, 120, 140, 194, 196–98
Gadow, Hans, 69–70
Galápagos archipelago, 47, 188
Galbreath, Kurt E., 163
Gallmaking and insects, 95, 97–99
Gastropoda, 170–71
Gauromydas genus, 143–48, 224
Gauromydas heros, 143–44
Gauromydas mateus, 143, 148
Gauromydas mystaceus, 144
Gauromydas papavero, 143, 145
GBIF (Global Bio-diversity Information Facility), xix
geckos, 5, 85–86, 88, 90–91
genera, 4, 97, 101, 112, 129, 136, 138, 143, 166, 172, 180, 187, 221
genitalia, 145, 156
genomes, 23, 26, 61–62, 72, 157, 174
genus, 27, 55, 57, 67–68, 70, 89, 112, 136–37, 144–45, 148, 179, 193–94, 219, 221–32
Geoffroy Saint-Hilaire, Étienne, 33
Geological Survey of India, 172
German beer mat nematode, 163–64
Germany, 9, 45, 69, 86, 89, 144, 186, 196, 223
Gertsch, Willis, 151–53
Ghana, xxi, 121, 135, 137, 139

Gibraltar, 213
Global Biodiversity Information
    Facility (GBIF), xix
Gobiesox lanceolatus, 218
Godwin-Austen, Henry Haversham,
    171–73, 175
Goethe University in Frankfurt, 9
Gold Coast, Africa, 137, 172
Goldsmith, Oliver, 122
Gorgosaurus, 204, 207
Grallaria ruficapilla, 105
Gray, George Robert, 45
Great Trigonometrical Survey of
    India, 171
Green River Formation, Colorado, 104
Guangdong Province, China, 131
Guiana, 35
Gymnodactylus marmoratus, 86
Gynacantha congolica, 119–20, 123

Haeckel, Ernst, 211
Hall, Sally, 186–30
Hamburg, Germany, 96
Hamilton, Chris A., 152–58, 224
Hanken, Jim, 67–73
Hanson, Michael, 205
Hartert, Ernst, 44
Hartop, Emily, 179
Harvard Museum of Comparative
    Zoology collection, 16, 67–68,
    75, 77–79, 218, 224
Haukisalmi, Voitto, 167
Hawaii, xvii, 127, 130–32–33
Heckel, Johann Jakob, 55
Helen, the brig, 51–54
Helgen, Kristofer M., 3–10, 25–26, 29,
    41–42, 47–49, 219
helminthology, 167
Hemidictya frondosa, xx
herbaria, xv, 195–98
herpetology, xv, xvii, 67–68, 75–77,
    81–82, 85–87, 90–31

Hershkovitz, Philip, 7
Highland mangabey lophocebus
Himalayan land snail, 169
Himalayas, 170–71, 174, 218
Hippocampus genus, 61
HMS Beagle, 111, 113, 218
Hoberg, Eric, 161–67, 224
Hoekstra, Paul H., 194–98
Holarctic, 166
Holland, 157
hominins, early, 212, 214
Homo erectus, 212–13, 220
Homo heidelbergensis, 214
Homo rhodesiensis, 214
Honolulu, Hawaii, xvii, 47, 127–28,
    132, 189
Hooker, Joseph, 113, 217–18
Hoolock tianxing, 219
Hope Entomological Collections,
    Oxford University, 115
Horn, Walther, 115
host-parasite relations, 161, 163, 165–
    67, 179
Hunyadi, András, 173
Hymenoptera, 97
Hypancistrus phantasma, 57

IBoL (International Bar-code of Life),
    24, 221
ichthyologists, xix, 52–53, 56
ICZN (International Commission on
    Zoological Nomenclature), 28–
    29
iDigBio, xix–xx
Igarapé Belmont, Brazil, 20, 26
Independent State of Papua New
    Guinea, 42
India, 47, 171–74, 220
Indonesia, 41–42, 47, 54, 86–88, 127,
    130–32, 211
Innuinnaqtun, 165
insect genitalia, 145–46

insects, xv–xvi, 54, 98, 100, 105–6, 113, 115–16, 120, 131, 133, 137, 144–45, 147, 152–33
institutions, xviii–xix, 6, 68–69, 79, 123, 145, 189
Instituto Oswaldo Cruz, Brazil, 145
Integrated Digitized Biocollections, xix, 105
International Barcode of Life (IBoL) project, 24, 221
International Commission on Zoological Nomenclature (ICZN), 28–29
International Union for Conservation of Nature (IUCN), 123, 175
intestines, 78, 161–63, 165
invasive species, xxi, 81, 175
Ipassa Research Station, Gabon, 194
Iquitos, Peru, 36
Irapuato, Mexico, 71
Irrawaddy River Delta, Myanmar, 171
Isangi, the Democratic Republic of the Congo, 121
isopods, 177–82, 190
isozymes, 70
Italian Natural History Museums, 220
IUCN (International Union for Conservation of Nature), 123, 175
Ivie, Wilton, 154
Ivindo National Park, Gabon, 194

Japan, 88, 169, 173, 185, 220
Java, Indonesia, 47, 90, 211–15, 220
Jones, Trevor, 27
Joordens, Jose, 211–15
Jordan, Karl, 44, 106
Josefin Stiller, 60–64
Juhel, Pierre, 137, 139, 224

Kaiser, Hinrich, 85–86, 88–91
Kansas, 128, 130
Kashmir, 171

Katanga junglewatcher, 123
Katanga longleg, 123
Kathirithamby, Jeyaraney, 133
Kays, Roland, 8–10
Kew Gardens, London, 16, 198
king crabs, 186–88
King Leopold II, 119
Kingsley, Mary, 135–37, 139–40
Kinshasa, Democratic Republic of Congo, 120, 124
Kipping, Jens, 120
kipunji, Rungwecebus monkey, 27
Kiska Island, Alaska, 181–82
Kogia breviceps, 162
Kraft, Richard, 36
Kullander, Sven O., 52–53, 55–57, 223
Kupang, West Timor, 88

labels, xv, xvii–xviii, 85–86, 99–100, 123–24, 128, 135, 139, 145, 147, 170, 174, 181, 186, 190
Laboratorio Nacional de Genómica para la Biodiversidad, Irapuato, Mexico, 71
Lambeosaurus, 204
land snails, 170–71, 173–74
Lascaux, France, 213, 220
Las Máquinas, Ecuador, 6
Las Vegas, Nevada, 151, 158
Lemaitre, Rafael, 188–90
Lenomyrmex hoelldobleri, xvi
Leopoldville, Belgian Congo, 120
lepidoptera, 76, 106
Lesser Karakorams, 171
Lesser Sunda Islands, 87
Lesson, René, 44
liana, 193–94
Linnaeus, Carl, 33, 197–98, 219
Lithodes couesi, 190
Lithodes galapagensis, 188
Lithodidae, 186–87, 190
Livingstone, David, 138

Lixophaga sphenophori, 131
Loader, Simon, 81–82
Logan, Utah, 96–97, 99–101, 221, 224
Lombok, Indonesia, 91
London, United Kingdom, xvii, 43–
44, 69, 85–87, 89–91, 96, 113–
15, 135, 138, 169–71, 217, 219,
224–33
longhorn beetles, xxi, 136–37, 139
Lophocebus kipunji, 27
Los Angeles, California, 86, 158, 177–
78, 182
Los Angeles County Museum of
Natural History, 86, 158, 177–
78, 182
Loveridge, Arthur, 78–79
lowland tapir, 20–23, 25–26, 28
Loyola Marymount University, 177
Lucihormetica luckae, 103
Lumbricus terrestris, 164
luminescence, 103–4

Madagascar, 95, 101
Makarov, V. V., 187
Makarov's King Crab, 185
Makovicky, Peter J., 205
malacology, 169, 171–73, 175
Malawi, 79
Malay archipelago, 54, 86
Malaysia, 42, 130
Maluku Islands, 45, 131
mammals, xvii, xix, 3–5, 7–8, 10, 13,
16, 20–23, 41, 43, 46–48, 164–
65, 206–8–29
Manaus, 28, 52, 55–56
Marine Biodiversity Center, 178
Markgraf, Georg, 157
Marsh, Laura K., 31–38
Marsh, Othniel Charles, 202
marsupials, 15, 48, 207
Mary Kingsley's Longhorn Beetle,
135–37, 139

Masherbrum, 172
Mayer, Alfred G., 23
Mayr, Ernst, 43–44, 46–48, 223–30
Megaselia, 179
Megaselia armstrongorum, 179
Megaselia mikejohnsoni, 179
Melin, Douglas, 55–57
Mendelson, Joseph R., 81
Menke, Harold W., 202, 225
Mermis nigrescens, 164
mesosoma, 97, 99
metasoma, 97, 99
Mexico, 70–71, 95, 157
Mexico and Central America, 154
Meyer, Adolf, 45
Mézière, Nicolas, 120
mice, 14, 16, 46
Microperoryctes aplini, 41, 43, 47–49,
223
Microperoryctes genus of bandicoots,
48
Minas Gerais, Brazil, 19
miniature tarantulas, 152
Minipteryx robusta, 146
Mojave Desert, 153
molecular analysis, 5, 23, 68, 196
mollusk collection, 169–70, 175, 212,
220, 224
Monanthotaxis barteri, 198
Monanthotaxis couvreurii, 194–95
Monanthotaxis genus, 193–94
Monanthotaxis latistamina, 195–96
Monanthotaxis specimens, 194–98
Monanthotaxis zenkeris, 196
Moniezia expansa, 167
monkeys, 27, 32–34, 37–38, 51, 54
Monodelphis sanctaerosae, 26
morphological characteristics, 8, 15,
20, 25, 35, 112, 135, 145, 153, 157,
173, 196
moths, xv, 145, 179, 219
Mount Arfak, 42

Mount Bonthain, 86
Mount Cameroon, 137, 140
Mount Everest, 172
Mount Lawu, 215
Mount Nimba, 76
Mount Rungwe, 27
mouse, 13–15, 17, 41, 43
    ashy-bellied Oldfield, 17
    beady-eyed Oldfield, 17
    strong-tailed Oldfield, 17
mouse bandicoot, 48
Mozambique, 97
Muir, Frederick, 127–29, 131–32
Munich, Germany, 36, 56, 223
Munro, Stephen, 212–13
Murie, Olaus, 161–63
Murielus harpespiculus, 162–63
museum, xvii, xix, 68–69, 75, 77, 88,
    90–91, 114–15, 119, 138–39, 144,
    196–97, 202–3, 223–24–32
museum collections, 4, 8, 15, 27, 32,
    36, 49, 90, 144–45, 155, 166, 174,
    181, 215, 219–20
    well-cataloged, 4, 96
Museum Koenig in Bonn, Germany,
    186
Museum National d'Histoire Natu-
    relle, Paris, 33, 147, 186, 196
Museu Paraense Emilio Goeldi, Bra-
    zil, 145
Mustela frenata, 7
Muztagh Glacier, 172
Myanmar, 171, 174, 219
Myoictis melas, 48

Nacional, 36
Nagano Prefecture, 173
Nairobi, 46, 96
National Herbarium of Cameroon,
    196
National Museum in Papua New
    Guinea, 47

National Parasite Collection, 162–63
National Science Foundation, xix
native collectors, 79, 104, 106, 172
Natterer, Johann, 55
natural history collections, xiv, xvi–
    xx, 32, 44, 48, 51, 55, 59, 89, 158,
    167, 175, 180–81, 218, 220–21
Naturalis Biodiversity Center, 195,
    211, 215
Naturalist's Color Guide, Smithe, 13
Naturmuseum Senckenberg in Frank-
    furt, 186
Neanderthals, 211, 213–14
Nematoda, 161–64, 218
Neodythemis katanga, 123
Nepal, 174
Netherlands, 194, 211
Neuroterus quercusbaccarum, 98
Nevada, 153
Nevada Test Site, 151–52
New Guinea and Papua New Guinea,
    41–49, 77, 131–32, 223–31
New Mexico, 101, 158
New York Botanical Garden Her-
    barium, 197
New York City, 36, 78, 223
next-generation sequencing, 157
Ngozi Crater, 78–79, 82
Nigeria, 79
nomina dubia, 155, 157
North Carolina Museum of Natural
    Sciences, 8
Notogomphus intermedius, 123
Nunavut, 164–65

Oaxaca, in southern Mexico, 68, 70–
    72
ocean, xxi, 42–43, 46, 54, 56, 63–65,
    163–64, 185
Ochotona cansus
Ochotona princeps, 161
Ockenden, George, 46

Odonata, 105–6, 120–21, 124
Ohbayashinema aspeira, 161, 163
Okhotsk, 165
Oldfield mice, 17
olingos, 3–6, 8–9
olinguito, 3–10, 25, 194, 223
Oophaga sylvatica, xvii
operculum, 170
Opuntia soederstromiana, 16
Orange River, South America, 140
Oredoxia regia, 54
Oriximiná, Brazil, 144
Orizaba, Mexico, 68–70
orphaned collections, xix, 189
O'Shea, Mark, 77
Oslo, Norway, 197
Otonga Reserve, 8–9, 193
Oxford University Museum of Natural History, 115

Pacific Ocean, 45, 61, 71, 175, 177, 179, 185
Palawan, Philippines, 188
paleobotany, xix
Paleolithic cave paintings, 213
paleontology, 19, 201–5, 207, 217
Páll-Gergely, Barna, 169–71, 173–75
Palmer Station, Antarctica, 165
palms, sago, 132
Panagrellus redivivus, 163
Pan-American Highway, 105
Papallacta, Ecuador, 16–17
Papavero, Nelson, 144
Papua New Guinea, 42, 45, 47, 131, 178
Paralithodes camtschaticus, 187
Paralomis bouvieri, 187
Paralomis makarovi, 185, 187, 190, 224
parasites, 98–99, 127–29, 131, 133, 161–67, 180
Pará State, Brazil, 29, 145
paratypes, 37, 63, 114, 224

Paris, France, xvii, 33, 44, 68, 96, 131, 147, 173, 186, 196–97, 218
Paseo del Mar, California, 177–79, 181
Patterson, Bruce, 24
Payden's isopod, 177
penis, 145–46
Pentcheff, Dean, 177–81
Peru, 15, 26, 35–37, 46, 56, 96, 105, 148
pests, agricultural, 98, 128
Phalloblaster, 146
Phascolosorex dorsalis, 48
Philippines, 54, 188–89
Phocoenoides dalli, 182
Phycodurus eques, 60
Phyllopteryx dewysea, 59
Phyllopteryx taeniolatus, 60
pika, 161–62, 165–66
Pink Floyd, 123
pipevine swallowtail, 69
Pithecanthropus erectus, 212
pithecia capillamentosa, 223
Pithecia genus, 31–35
Pithecia monachus, 33
Pithecia pissinattii, 34
Pithecia rylandsi, 34
Pithecia vanzolinii, 37–38
Placentonema gigantissima, 164
plants, xv, 37, 98, 122, 155, 164, 193–96, 198
Pocock, Reginald, 153
Pontia daplidice, 115
Port of Los Angeles, 178
Porto Velho, Brazil, 20–21
Poupart, François, 121
Prague Insect Fair, 137
Pseudictator kingsleyae, 135–37, 224
Pseudodon vondembuschianus trinilensis, 211, 213
Pseudopomatias abletti, 169–70
Pseudopomatias genus, 170
Pseudopomatias himalayae, 170, 172

Pseudopomatias reischuetzi, 175
Pseudoricia flavizoma, 219
Pyrophorus beetles, 104

Quercus robur, 98
Quichua language, 107
Quindío Province, Colombia, 105
Quito, Ecuador, 15–17, 35, 106

rainforest, xxi, 19–20, 28–29, 34, 36, 42, 51–52, 87, 104, 130, 193–94, 197–98
Rana kauffeldi, 78
Recherche archipelago, Australia, 60, 62, 65
Recife, Brazil, 157–58
Red Deer Badlands, 203–4
Red List of Threatened Species, 123
reptiles, 37, 43, 69, 75, 77–78, 86, 106, 112, 137
revision, 37, 112–14, 170, 174, 195
Rhabdoscelus obscurus, 127
Rhipidocyrtus muiri, 127–28, 130, 132, 224
Rhoads, Samuel N., 16
Riggs, Elmer S., 201–7, 225
Riley, Charles, 131
Rio Madeira, 21, 26
Rio Negro, 35, 52–53, 55–57
Rio Papallacta, 14
Rio Sepotuba, 22
Ripidiini beetles, 128–29, 133
Ripiphoridae, 129
river, 20–22, 29, 34–35, 55, 121, 164, 211–12
River Amazon, 35
River of Tapirs, 22
Rodentia, 14, 17, 207
roly-poly, marine, 177–78, 180–81
Roman, Abraham, 55
Rondon, Colonel Cândido, 21
Rondônia State, Brazil, 19–20, 29

Roosevelt, Theodore, 21–23, 29, 33, 223
Roosevelt River, Brazil, 29
Rothschild, Lord Lionel Walter, 43–44, 46
roundworms, 162–63
Rouse, Greg W., 60–61, 63–64
rove beetles, xiv, 111–13, 115
Rovito, Sean M., 71–72
Royal Belgian Institute, 120
Royal Botanic Garden Edinburgh, 89
Royal Holloway, University of London, 217
Royal Museum for Central Africa, Tervuren, Belgium, 119–20, 122, 224
ruby seadragon, 59, 62–65, 223
Rudolphi, Karl, 167
Russia, 185

Saguaro National Park, Arizona, 153, 155, 158
saki monkeys, 31–37
salamanders, 68–69, 71–72, 80
lungless pygmy, 67, 219
samples, xvii, 61–62, 101, 108, 163, 165, 171, 224
San Agustín, Colombia, 7
Sandwich Islands, 130
San Miguel Suchixtepec, Mexico, 72
Santa Catalina Mountains, 158
Santo Tomás Teipan, Mexico, 71
Saskatchewan, Canada, 166
scales, xv, 56–57, 77, 90, 147, 221
Schmidt, Justin O., 147
Schmidt Sting Pain Index, 147
scientists, xvii, xx, 24, 27, 32, 44, 53, 60, 87, 122, 144, 146, 171, 189, 219–20
Sciurus nayaritensis chiricahuae, 155
Scottnema lindsayae, 163
Scripps Institution of Oceanography, San Diego, 60, 218

seahorses, 61
Sedaris, David, 114
Semioptera wallaci, 45
Senckenberg Museum, Frankfurt, 90, 173
serrate antennae, 112
Seycr, Charles Henri, 123
shelf life, xviii
shells, 46, 137, 169–70, 172–75, 211–15, 225
shrews, 41, 48, 165
Siberia, 165, 214
Sierra Leone, 136, 139, 156, 198
Sierra Madre, 71, 154
Simpson, Bill, 201–8
Sirdavidia solanonna, 198
Sixth Extinction, xx
skins, 3, 5, 21–23, 32, 36, 46–47, 52, 61, 82, 85, 173
skulls, 3–6, 19–23, 27–28, 32–33, 36, 47–48, 155, 196, 205, 207, 212
Skywalker hoolock gibbon, 219
Smith, Andrew, 157
Smith, Malcolm A., 86
Smith's hidden gecko, 85
Smithsonian Institution, xvii, 90, 130, 139, 180, 182, 186, 188, 195, 218–35
snails, 172–73, 175
snakes, xv, 21, 69, 77, 85–87, 90, 105, 173
Snow, Sally, 186–88, 190
Snow Mountains, New Guinea, 49
Soderstrom, Ludovic, 15–17
Solodovnikov, Alexey, 114
Sorkin, Louis, 153
Soucoupe, submersible, 219
South Africa, 97, 120, 140, 218
South America, 3, 20, 24–25, 33, 35, 52–53, 61, 103, 206–7
Southern Conquest, research vessel, 65

speciation, 54
Sphaeromatidae, 180
Sphenophorus genus, 130, 132
spiders, xviii, 42, 56, 105, 151–53, 155–58
Spix, Johann Baptist, 34, 223
Spruce, Richard, 52
squeaker frogs, 75–76, 78, 80, 124
Staphylinidae, 112
Star Mountains of New Guinea, 77
Stauroteuthis albatrossi, 190
Stegosaurus, 203
Stellenbosch University, South Africa, 120
Stiller, Josefin, 60–64
Stimpson, William, 179–80
Stockholm, Sweden, 52, 56–57, 96, 223
Sturnira lilium, 24
Sus papuensis, 46
Sweden, 34, 52–53, 55–56, 167, 197–98, 223
sympatric, 70
Syngnathidae family, 59

Tan, Milton, 57
Tangara pulcherrima, 16
Tanzania, 27, 77, 80, 82
tapeworm, 166–67
tapirs, 19–23, 25–29, 218, 223–29
Tapirus kabomani, 19–21, 23, 26
Tapirus pinchaque, 20
Tapirus pygmaeus, 28
Tapirus terrestris, 20, 26
tarantula hawk wasps, 147
tarantulas, 147, 151–54, 156–58
Tate, George Henry Hamilton, 6
taxonomic impediment, xx, 88, 101
taxonomists, xv–xvi, xviii, 23–24, 27–28, 32–34, 112, 114, 138–39, 145, 166, 170–71, 177, 179, 189–90, 220–21

Teton Range, Wyoming, 161
Texas, 42, 96, 101, 162, 205
Thelbunus mirabilis, 146
Theraphosa blondi, 158
Thomasomys genus, 16–17
Thomasomys ucucha, 13–14, 223
Thorius arboreus, 67
Thorius genus, 67–73
Thorius grandis, 71
Thorius lunaris, 70
Thorius magdougalli, 71
Thorius magnipes, 70
Thorius minutissimus, 71–72
Thorius papaloae, 71
Thorius pennatulus, 68–69
Thorius pinicola, 72
Thorius pulmonaris, 68, 224
Thorius schmidti, 71
Thorius spilogaster, 70
Thylacosmilus atrox, 207
Tillack, Frank, 82
Timor, 86–89, 91
Tipulidae, 105–6
Tiputini Biodiversity Station, 31–32
toads, 81, 87, 105
Tomales Bay, California, 181
Tomiya, Susumu, 205
Torvosaurus, 205–6
Toxicocalamus genus, 77
Trans-Mexican Volcanic Belt, 70
tribes, xix, 26, 37, 42, 112, 128, 138–39
Triceratops, 201–3, 207
Trigonopselaphus genus, 112–14
Trinidad, 148
Trinil, 211–12, 215
Tropical Queensland, 180
tundra, 165, 167
Tungurahua Province, 104–5, 107–8
Tweedie, Ethel, 136
type specimens, xvii, 6, 24, 33–34, 36, 90, 138, 155, 162–63, 169, 171
Tyrannosaurus, 204

ucucha mouse, 13–15, 17, 25
Uganda, 79, 120
Universidade Federal de Minas Gerais, 19
Universidad Federal de Rondônia, 20
Universidad Nacional de la Amazonía Peruan, 36
University of Michigan, 13–15, 224
Upper Rio Negro in Brazil, 52
Uppsala, Sweden, 197–98
Ursus arctos, 9
Ursus maritimus, 9
USS *Albatross*, 185, 188–90–35
Utah, 96, 99–100, 203, 221, 224

Van Roosmalen, 28–29
Venezuela, 25, 220
Victor Valley College, California, 86
Vietnamese land snail, xxi
Vilars, Arthur, 55
volcano, xxi, 16, 42, 70, 78, 106–8, 163, 215
Voss, Robert S., 13–15, 17, 25–26, 29
voucher specimens, 25, 27, 169
Vršanský, Peter, 103–4, 107–8

Wageningen University, Netherlands, 194
Wake, David, 68
Wall, Adam R., 178, 224
Wallace, Alfred Russel, 34–35, 42, 45, 51–57, 155–30
Wallacea, 54
Wallace Line, 54
Wallace's Pike Cichlid, 51–52, 57, 223
Wallace's standardwing, 45
WAM (Western Australian Museum), 59–60, 62, 64, 223
Washington, DC, xv, xix, 7, 68, 95–96, 101, 103–4, 130, 146, 148, 161, 181, 186, 224

wasps, xx, 95, 97–99, 101, 138, 146–48

weevil, 127–28, 131–32, 145–46, 218

West Africa, 135–37, 140

Western Australia, 59–60, 64

Western Australian Museum. *See* WAM

Western Cape, South Africa, 99 100, 218

West Kalimantan, Borneo, 127

West Timor, 87–88

Wetzer, Regina, 178, 180

Wheeler, Quentin D., xvi, 190

William Healey Dall, 181, 224

Wilson, Nerida, 60–61, 64–65

wings, 5, 97, 103, 115, 120, 122, 125, 129, 144, 148, 205, 219

worms, xiv, 17, 161–67, 218–19

Wyoming, 99, 162–63, 165, 203, 205

Xanthopygina, 112

Yale Peabody Museum of Natural History, 203

Yaoundé, Cameroon, 193

Yule Island, 45

Zambia, 123–24

Zenker, Georg August, 196–97

ZooKeys, journal, 96, 114, 224–29–34

Zoologische Staatssammlung, Munich, 36, 223

Zootaxa, journal, 170, 188–36